Opening the High Frontier

Our Future in Space

Written by Eagle Sarmont

Graphics by Nick Stevens

Triplanetary Publications, San Diego CA 2016

Published by:

Triplanetary Publications
P.O. Box 601433
San Diego, CA 92160.

Soft cover: ISBN 978-0-692-76002-4
Printed in the United States of America

Photo Credits:

NASA and U.S. Air Force photographs and graphics available as public domain.

Historic artwork and photographs are available as public domain.

Available as Creative Commons:

The photo of SpaceShip One on page 43 is by D Ramey Logan.

The graphic of the Space Elevator on page 59 is by Skyway and Booyabazooka.

The photo of the V-1 on page 62 is by Nick-D, May 2006.

Published with written permission:

The Front and Back Covers, interior graphics of the Skyhook, Umbrella Ship, Martian Electric Spaceship, Kuck Mosquito, and Space Colony near Earth are by Nick Stevens.

The interior graphic of the Sun Ship at Jupiter is the result of a collaborative effort by Graham Gazzard and Nick Stevens.

Digital scans of paintings by Chesley Bonestell: *Assembly of the Moonships 1,075 miles above the Hawaiian Islands, 1952*; and, *Approaching Mars, 1948*; are reproduced courtesy of Bonestell LLC.

Digital rendering of the Stratolaunch carrier aircraft is courtesy of Vulcan Inc.

Photo of New Shepard space vehicle is courtesy of Blue Origin.

The number of people who have knowingly or unknowingly contributed to this book is very large. We all stand on the shoulders of those who came before us.

Thank you to all.

For all the dreamers and the doers of the world,

past, present, and future.

Space tourism, space hotels, orbiting factories, outpost space stations, asteroid mining, mining the Moon, Space Colonies, satellite solar power stations, cities on the Moon and Mars, spaceships that travel from the Earth to the Moon, to Mars and beyond, people living and working in outer space, these are the dreams of mankind about space.

For these dreams to become reality, we need Earth to orbit transportation, and Earth orbit to escape velocity transportation, that just about anyone can afford.

This book is about how we can make this happen using existing technology, and accomplish it affordably.

It's about the ideas and technologies that will allow us to become a spacefaring civilization.

Exploring the Moon in 1972, Apollo 17, NASA

Dare greatly.

Learn from your mistakes.

Don't stop.

Contents

This view of the non-rotating Skyhook shows the Central Space Station moving down the cable towards the Lower Endpoint Station.   By Nick Stevens.

Foreword

Hello, my name is Eagle Sarmont and I want to speak with you about some wonderful and incredible things: about manned space exploration; about cities on other worlds; and, about space colonies scattered throughout the solar system where people will live, work, and raise families. In short, I want to talk with you about mankind becoming a spacefaring species, about our building a spacefaring civilization with orbiting hotels, orbiting factories, asteroid mining, satellite solar power stations, and space colonies, about terraforming Mars and filling the solar system with life, and someday going to the stars.

I call it *Opening the High Frontier* and it's about how we can make spaceflight affordable to just about everyone.

- If you could go to a spaceport and buy a ticket to orbit and back for $20,000, would you do it?
- If you could get a job on a Space Colony and move there, would you go?
- If you could sign up as a crew member on a ship that was going asteroid mining where you had the chance to come home a multi-millionaire, would you go?
- If you could get a job working on a Satellite Solar Power Station in space, would you take it?
- Everyone has heard about global warming. Did you know that Satellite Solar Power Stations built from asteroid or lunar materials would allow us to significantly reduce our contribution to it? They would.
- Everyone knows about our over-population problem. Did you know that it is possible to build land area in Space Colonies at a faster pace than our population is growing? It's true.

In short, outer space is about solutions and opportunities, it's about jobs and earning a living, it's about building a new way of life, and it's about solving our problems on Earth.

The only thing holding us back is figuring out how to make spaceflight affordable so we can start solving those problems.

The movies make spaceflight look easy with anti-gravity motors, warp drives, and matter transmitters. Unfortunately, we don't know how to make any of those things. Unlike Hollywood, this is the real world and we need to figure out how to make spaceflight affordable using what we know how to build.

That is what this book is about, how to make spaceflight affordable to just about everyone, using existing technology, using known ideas, and how to build it for a price we can afford.

Boomerang Nebula, Hubble Space Telescope

Saturn as viewed by the Cassini spacecraft, NASA photo

Prologue

Two possibilities exist: either we are alone in the universe or we are not. Both are equally terrifying.

Arthur C. Clarke

A long time ago, in a faraway solar system, a civilization discovered the ideas put forth in this book and used them to become a spacefaring people. They were not the first to do this, nor will they be the last. Gradually, over time, their civilization grew until it filled their solar system all the way out to the outer most fringes. At that point, some of them decided to take up the challenge of the stars and they built giant ships that were capable of surviving the long voyages. When they arrived at a new star system inevitably some of them wanted to stay and settle and some wanted to resupply and continue the adventure. As might be expected, both occurred.

As time went by, the star travelers grew better and better at building their giant ships so they could make the crossings between the stars in less time. As they grew in number, they also built additional ships so that they could either travel together in convoy or spread out to investigate nearby star systems along their general flight path.

On rare occasions, they also ran into other intelligent life forms with technological civilizations. Some of these races were star travelers like themselves, and some were content to stay where they were, but liked to interact with the various star traveling races. Gradually, as these interactions grew in number, an informal galactic community developed that involved the sharing of knowledge, experience, and culture. Some of the races even joined together to build really large space habitats where they might try living together to see what kind of joint community might evolve. Others became farmers of intelligent life and took to traveling a loop among stars that had planets that were capable of supporting life.

One thing they all looked for were emerging technological civilizations that were on the cusp of becoming a spacefaring people as these were prospective new members for their growing galactic community.

Is this speculation or truth?

No one on Earth knows for sure. According to the Drake Equation and Fermi's Paradox, the odds are greatly in favor of it being true. If that is the case, the question then becomes; are we one of the first to become spacefaring in this part of the galaxy, bound to become one of the founders of a galactic community, or are we one of the new species that has been unknowingly helped along?

Assembly of the Moonships 1,075 miles above the Hawaiian Islands, 1952

This painting by Chesley Bonestell depicts many of the key components of Wernher von Braun's vision for our future in space. The wheel-shaped space station shown in the upper left corner of the picture was spun in order to create artificial gravity. The two reusable winged space shuttles were for carrying cargo and crew from the surface of the Earth to orbit and back. They were launched using reusable two-stage rockets. In the foreground are two Moonships being prepared for a voyage to the Moon.

While very different in appearance from the actual Space Shuttles, space stations, Apollo spacecraft, and Lunar Landers that have since been built, there are still many worthwhile ideas contained in this picture that have yet to be realized.

i Wernher von Braun's Vision

From 1951 to 1957, Dr. Wernher von Braun, with the assistance of Collier's Magazine, Willy Ley, Walt Disney, Chesley Bonestell, and others, shared his vision of our future in space with the world. It was an astonishing and wonderful vision that captured the imagination of just about everyone who heard or read about it. It was a vision that included fleets of large multi-stage rockets, reusable space shuttles, space stations in orbit around the Earth, space telescopes, spacecraft for traveling from Earth orbit to the Moon and back, and a fleet of interplanetary spacecraft for a manned expedition to Mars. It is a vision that has guided the space programs of the world ever since.

Now, over 60 years later, we have built the multi-stage rockets, the space shuttles, and the spacecraft. We have walked on the Moon. We have built space stations and space telescopes. The only part of this vision we have not fulfilled is to send a manned expedition to Mars.

Dr. von Braun and his vision changed the world. From the very beginning, Dr. von Braun's vision was seen as a large government funded program. This has been both its strength and its weakness. It was strength in that it got us to the Moon 12 years after the first satellite was placed in Earth orbit. It was a weakness, like most government funded programs, that no attention was given to making it affordable; and, as a result, we have not progressed very much since.

In the beginning, the four primary reasons that supported and justified the space program in the United States were:

1) to seize the high ground before the Soviets did;
2) to demonstrate to the rest of the world that capitalism was superior to communism;
3) to stimulate technological development in this country so as to maintain our technological superiority; and,
4) to open the space frontier for our children and our childrens' children.

Towards the end of the 1960s, the first three reasons had been resolved and it was only the last one that really mattered to anyone anymore. Unfortunately, that dream was put on hold when the $300 million cost per Moon flight became common knowledge. After all, how many people would ever be able to afford a trip to the Moon when the round trip ticket price is $100 million per person? It was at this point that Dr. Wernher von Braun's vision for opening the space frontier for our children and our childrens' children became an unrealistic and unrealizable fantasy for most people.

Stanford Torus *D. Davis, NASA*

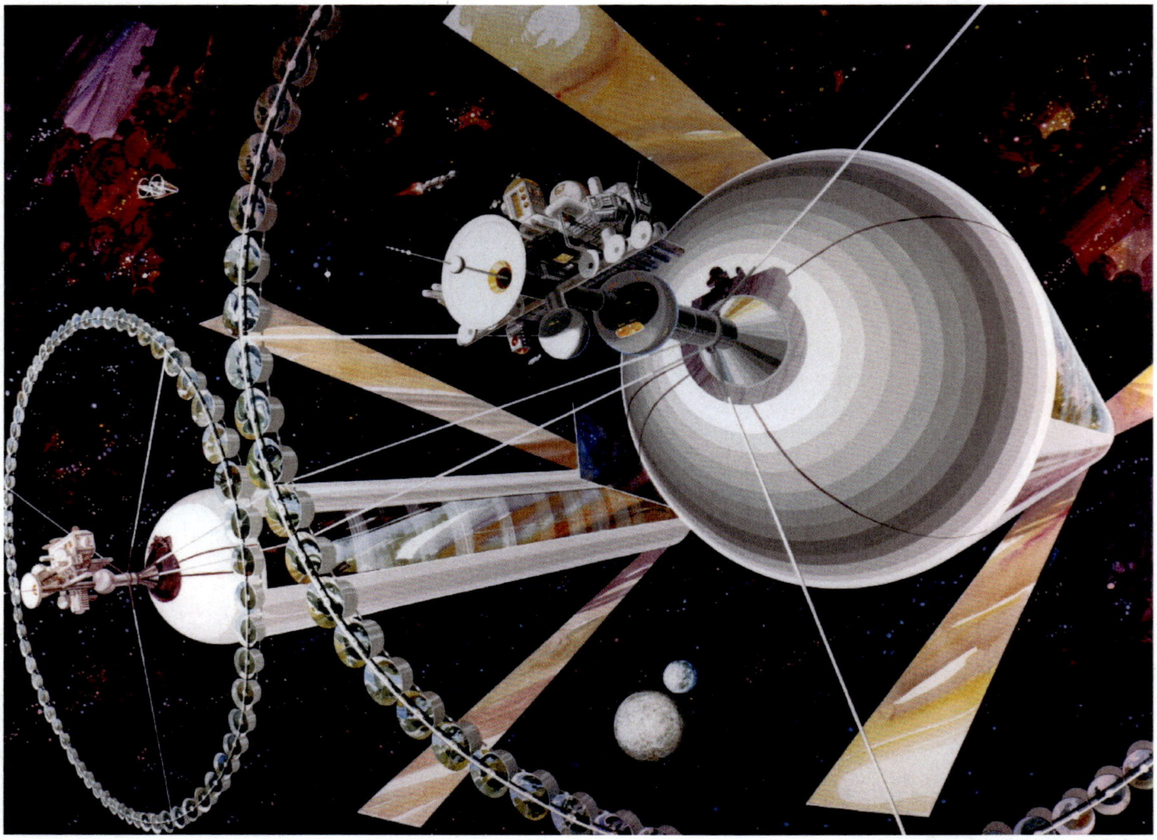

R. Guidice, NASA

ii "Islands in Space"

In 1977, Dr. Gerard K. O'Neill published a book about his vision for our future in space called *The High Frontier: Human Colonies in Space*. It was a vision that included giant self-contained cities in space called "space colonies" or "islands in space", as well as "mass drivers" on the Moon for launching lunar materials into space for building those cities. The purpose of these cities was to build giant space-based solar power stations for beaming affordable, pollution-free power to the Earth, and for building additional "islands in space" for our expanding population. In Dr. O'Neill's vision, outer space was not just something to be crossed on our way to another planet, it was the location and future home of our civilization. A civilization that he thought would grow until it becomes many thousands of "islands in space" spread throughout the solar system.

The foundation that this vision was built upon was the planned existence of affordable Earth to orbit space transportation, transportation that was to be provided by the then yet to fly Space Shuttle. At that time, the Space Shuttle was supposed to be able to carry approximately 50,000 pounds into low Earth orbit at a cost of $25 million per flight ($500 per pound). Over time, this cost was expected to drop to $100 per pound as the flight rate increased and as we got better at building space shuttles. Unfortunately, the Space Shuttle ended up costing over $1 billion per flight. A cost that made Dr. O'Neill's incredible vision into an unaffordable dream.

R. Guidice, NASA

"Hands amidst the stars" *NASA photo*

iii Opening the High Frontier

All this world is heavy with the promise of greater things, and a day will come, one day in the unending succession of days, when beings, beings who are now latent in our thoughts and hidden in our loins, shall stand upon this Earth as one stands upon a footstool, and laugh and reach out their hands amidst the stars.

H. G. Wells, 1903

The ideas and technologies contained in this book will enable the affordable movement of people and goods from the surface of the Earth to orbit and beyond. They will make spaceflight affordable to just about everyone on a commercial basis. It is these ideas and technologies that will allow mankind to successfully explore, develop, and colonize this solar system and make real the visions of Dr. Wernher von Braun and Dr. Gerard O'Neill. It is about opening the high frontier for ourselves, our children, and our childrens' children.

It is about Our Future in Space.

NASA photo

An Orion spacecraft approaching the lower end of the Skyhook
By Nick Stevens

Chapter 1: The Known Past

The farther backward you can look, the farther forward you are likely to see.

Winston Churchill

Approximately 195,000 years BCE (Before Common Era), early modern humans began to evolve in northeastern Africa in an area now known as Ethiopia. It is an evolution that has continued ever since. Current data suggests these early modern humans evolved from archaic humans, who had evolved from Homo heidelbergensis, who had evolved from Homo erectus, who had evolved from Homo habilis.

Approximately 80,000 years later, or 115,000 BCE, early modern humans had expanded their range to include most of Africa as well as a small part of southwest Asia now known as Israel. At this time, there is no reliable evidence of early modern humans in other parts of Asia until 60,000 to 40,000 years BCE.

There were very few early modern humans living during this period. At one point, shortly after the Toba catastrophe of 73,000 BCE, the total human population is estimated to have dropped to somewhere between 1,000 and 10,000 breeding pairs. With numbers this small scattered over such a large area, it is no wonder that reliable evidence is scarce.

It was sometime after this when early modern humans began moving further into Asia and into Europe, that they came into contact with older groups of archaic humans and Neanderthals who had migrated out of Africa before early modern humans came into existence. Today, we have genetic evidence that shows these early modern humans interbred with both the archaic humans and the Neanderthals they met, creating a hybrid that is now known as modern humans. This hybridization process occurred primarily in the areas now known as Europe and China. The proof of this exists in the DNA of modern-day Europeans, which contains approximately 3.2% Neanderthal DNA, and the DNA of modern-day Chinese, which contains approximately 2.6%. It is also interesting to note that it is these two groups who have been the creators and builders of modern human civilization. In support of this observation, there is currently no known evidence that un-hybridized Neanderthals, archaic humans, or early modern humans ever progressed beyond primitive tribal life. So it would appear that these early modern human/Neanderthal hybrids, aka modern humans, are an example of the result being more than the sum of its parts.

Modern humans are still evolving. Natural selection, changing environment, competition, and survival of the fittest are still forces that drive us and shape us whether we like it or not. We have no control over most of these forces, but there is a growing body of evidence that shows that life experience and the choices we make as individuals can also change our personal DNA during the course of our lifetimes. This was first noticed in studies of identical twins that were separated at birth and who grew up and matured in different environments. At birth, the DNA of the twins was identical, but it changed over time in different ways depending on the level of stress they experienced. These levels of stress were due either to external events or to the choices they made and the challenges they decided to take on.

These choices that affect our evolution occur on both an individual level and on a group level. When an individual chooses to pursue an easy career path versus a more difficult one, they are limiting how much they will achieve, how much they will earn, what they will become, the opportunities they will have, and how far they will evolve. The same thing happens when a group or a civilization makes a similar choice.

In 1961, President Kennedy set us on course for going to the Moon.

We choose to go to the Moon in this decade and do the other things, not because they are easy, but because they are hard, because the goal will serve to organize and measure the best of our energies and skills, because that challenge is one that we are willing to accept, one that we are unwilling to postpone, and one which we intend to win, and the others, too.

John F. Kennedy

President Kennedy chose to go to the Moon because it was hard, not easy. By choosing the difficult path he challenged all of us as a nation and as individuals. The results of that choice are very clear; all you have to do is look at how we have evolved as a civilization and as individuals ever since.

NASA photo

Chapter 2: Transformational Events over the last 600 Years

We are not here merely to make a living, but to enrich the world with a finer spirit of hope and achievement, and we impoverish ourselves if we forget the errand.

Woodrow Wilson

1400 Total human population is estimated to be 350 to 374 million people.

1410 Up to the early 15th century A.D., Chinese technology led the world. Fleets of "treasure ships", some as long as 400 feet and weighing 1,500 tons, armed with cannon, navigated by compass and stars, sailed the seas of the Far East and the Indian Ocean. They sailed around Southeast Asia to Sumatra, to Bengal, to Ceylon, to India, and the Middle East. In 1418, they even sailed down the east coast of Africa, 70 years before the Portuguese sailed down its west coast to the Cape of Good Hope. It is possible that they also sailed around the southern tip of Africa into the Atlantic Ocean and part way up the west coast. Others have suggested that they also crossed the Pacific Ocean and visited the west coasts of North and South America. It was these "treasure ships" that made possible a trade network that brought home the wealth and the ideas of the world to China; wealth and ideas that made China the global superpower of the day.

The Treasure Fleet as depicted on an ancient Chinese woodblock print

Opinion is the main thing which does good or harm in the world. It is our false opinions of things which ruin us.

Marcus Aurelius Antonius

China made its last great voyage between the years of 1431 and 1433. In 1436, when a new Emperor came to the throne, an edict was issued that forbade the building of ships for overseas voyages. A social shift had occurred in China. The Confucius philosophy, that disliked innovation and change, that viewed traders, merchants, and business people as parasites on society, and viewed the life of the small family farmer living in tune with the Earth as the highest ideal that a person could aspire to, had gained control of society and the government. Desiring to become a kinder, gentler, and more stable society, China, under the Confucians, gave up the world and retreated into isolation. In effect, they gave up the challenging path and took the easy path. It was a decision that cost China its wealth, its leadership in technology, and its status as the superpower of its day. It was a decision that resulted in China's downfall and complete subjugation by the western world four centuries later.

We need to remember that we are not here on this Earth just to be users. We were made in the image of the Creator, and we are here to create. For us to create we must first acknowledge and explore the limits of our mind and soul: limits that must be expanded before we can expand any physical ones. So it is as with any endeavor, that the greatest distance to be crossed is that which lies within us.

Walter J. Hickel, former governor of Alaska

1418 Prince Henry the Navigator starts the first school of oceanic navigation at Sagres, Portugal.

1420 One of Prince Henry's sailing expeditions discovers the islands of Porto Santo and Madeira.

1427 Another of Prince Henry's expeditions discovers the Azores.

1431 Prince Henry founds what will later be known as the University of Lisbon.

1434 Prince Henry the Navigator convinces Gil Eannes to take a ship south down the west coast of Africa into the "Sea of Darkness" and go further than any European has gone before. It is the beginning of the Great Age of Discovery.

1450 Johannes Gutenberg invents the printing press.

1453 The fall of Constantinople to the Ottomans marks the end of the Roman Empire and the end of the Middle Age. Byzantine scholars fleeing the fall of Constantinople arrive in Florence, Italy, and under the patronage of the Medici help to start the Renaissance.

1480 King Ferdinand and Queen Isabella initiate the Spanish Inquisition in order to purge Spain of Muslims and Jews. By 1530, approximately 2,000 people had been killed and somewhere between 40,000 and 80,000 had left the country. The most intense period of the Inquisition was from 1480 until 1530, but it continued at varying levels until repealed in 1834. While the total number of people killed by the Inquisition for this entire period is believed to be less than 5,000 people, the total number of people who left the country is believed to be somewhere between 300,000 to 800,000 people. The recorded opinions of many people who lived during that time state that the Inquisition was designed to rob rich people. Considering how those who left were only allowed to take what they could carry (no jewels, gold or other valuables allowed), and how those who died took even less, it would appear that there was some truth to this.

The most immediate impact of the Inquisition was how it drove both educated people, and people of proven business ability from the country. Many of these people moved to Italy, further helping to fuel the Renaissance of knowledge and thought that was occurring there. But in addition to the long-term negative consequences of this outward migration of brains and productive talent, there was also the ongoing repression of thought, of new ideas, and the suppression of progress that occurred for those who stayed. In many ways, the long-term effects of the Spanish Inquisition on Spain were similar to what the Confucians had on China.

1488 Bartolomeu Dias sails down the west coast of Africa to the Cape of Good Hope and discovers the Indian Ocean.

1492 Christopher Columbus sails west and discovers the Americas.

1498 Vasco da Gama reaches India by sailing around the Cape of Good Hope and crossing the Indian Ocean.

1500 Total human population is estimated to be 425 to 500 million people.

1519-1521 The sailing ship *Victoria* becomes the first ship to sail around the world and Ferdinand Magellan becomes the first man to sail around the world.

Sail forth – steer for the deep water only,
Reckless O soul, exploring, I with thee, and thou with me,
For we are bound where mariner has not yet dared to go,
And we will risk the ship, ourselves, and all.

Walt Whitman

1543 Copernicus publishes his theory that the Earth and all the other planets orbit the Sun. This was in opposition to the prevailing view of the time that the Earth was the center of the universe and everything revolved around it.

1568 Bottled beer is invented in London.

1569 Gerard Mercator invents the Mercator map.

1588 The English Navy defeats the Spanish Armada. Spain sent a fleet of 130 ships to escort an army across the English Chanel that was to invade England. A significant factor that contributed to the defeat was Spain's failure to keep up with the latest developments in naval technology and naval warfare, developments that were being spearheaded by the English.

1600 Total human population is estimated to be 545 to 579 million people.

1607 Jamestown, the first permanent English colony in America, is founded.

1608 Hans Lipperhey applies for a patent for his invention of the telescope.

1609 Johannes Kepler publishes the first two laws of planetary motion.

1610 Galileo Galilei discovers four moons moving around the planet Jupiter using a telescope.

1618 Johannes Kepler publishes the third law of planetary motion.

1687 Isaac Newton publishes the law of universal gravitation.

1700 Total human population is estimated to be 600 to 679 million people.

1752 Ben Franklin goes kite flying in the rain and proves that lightning is made of electricity.

Lithograph of Franklin and his kite experiment by Currier and Ives, 1867.

1757 John Campbell invents the sextant.

1761 John Harrison invents the marine chronometer for measuring longitude.

1776 Thomas Jefferson writes the *Declaration of Independence* based on a philosophy of personal freedom, personal responsibility, the right to pursue your dreams and become more than you now are, freedom of speech, equality, justice, liberty, fraternity, the rights of man, democracy, and representative government. With this document, the American colonies declare their independence from the British Empire and start down the path of becoming the United States of America.

1781 James Watt patents a design for a steam engine that produces continuous rotating motion.

Side view of the Albion Mill steam engine designed by James Watt

1783 Man leaves the surface of the Earth for the first time in an untethered hot air balloon.

The Montgolfier brothers launch the first manned flight of an untethered hot air balloon.

1798 Reverend Thomas Malthus predicts a dark future for mankind based on the idea of what is now known as the limits to growth.

> *The power of population is indefinitely greater than the power in the earth to produce subsistence for man. That the superior power of population is repressed, and the actual population kept equal to the means of subsistence, by misery and vice.*
>
> *Reverend Thomas Malthus*

The Reverend wrote this in opposition to the popular view of his time that man was perfectible. As a man of the cloth, he saw this limit to growth as divinely imposed so as to teach virtuous behavior. Critics of the Reverend have often wondered how a man of God could propose that misery and vice would teach virtuous behavior, but that was a question he never answered.

This dark future and its attending philosophy of misery and vice, of holding down those beneath you, and hopelessness about the human condition and the future, had a profoundly negative effect on European society.

1800 Total human population is estimated to be 900 to 978 million people.

1801 Lord Dundas funds the development of the first practical steamboat.

Designed to tow barges in the canal system around Glasgow, the "Charlotte Dundas" was the world's first practical steamboat.

1803 Richard Trevithick builds the first railroad steam locomotive.

c. 1803

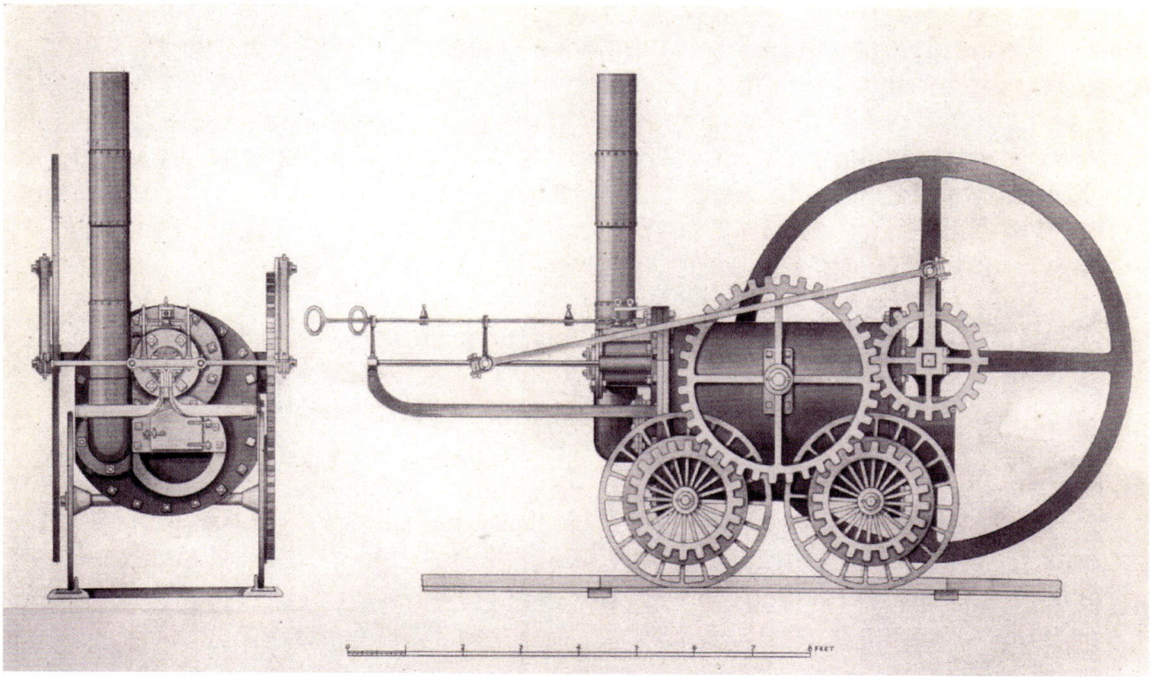

1819 The steamship *Savannah* makes the first steamship crossing of the Atlantic Ocean.

1835 Samuel Morse invents the telegraph.

1848 Karl Marx publishes *The Communist Manifesto*, a book about his vision regarding the inevitability of the formation of a communist state, about the conflict between the owners of the machines of production and the labor class that operates them, about an individual's responsibility to the group and the need for the individual to conform to the group, and the need for a centralized government to direct it – the "dictatorship of the proletariat" as he called it. One of the key tenants of this philosophy is contained in the statement,

From each according to his abilities, to each according to his needs.

Critics of this philosophy see this "worker's paradise" as a place where the people-of-ability would be enslaved by the people-of-no-ability and their never ending needs.

Karl Marx is also the inventor of "political correctness", or what critics have called "government dictated thoughts and opinions" designed to tell people what to think, what to do, and what to say.

Socialism is a philosophy of failure, the creed of ignorance, and the gospel of envy. Its inherent virtue is the equal sharing of misery.

Winston Churchill

1856 Henry Bessemer patents the first cost-effective industrial process for the mass-production of steel.

1861 The Western Union Company builds the first transcontinental telegraph.

1861-1865 A civil war in the United States of America brings about the end of slavery in that country. The population of the United States, as determined by the 1860 census, was 31,443,321 people, of which 3,953,761 were slaves. Total death toll for the war is estimated to be 620,000 people: approximately 420,000 people from the North; and, 200,000 people from the South.

1869 The first transcontinental railroad is built.

1876 Alexander Graham Bell patents the telephone.

1880 Thomas Edison patents the first practical incandescent electric lamp.

1889 The Eiffel Tower is completed in Paris, France. It is the tallest man-made structure ever built until the Chrysler Building is built in New York City in 1930.

1895 In response to the dark view of life and the future that was started by Reverend Malthus and his limits to growth philosophy, Konstantin Tsiolkovsky, the founding father of rocketry and astronautics, said,

The murky views which some scientists advocate as to the inevitable end of every living thing on Earth . . . should not be regarded as axiomatic. The finer part of mankind will, in all likelihood, never perish – they will migrate from sun to sun as they go out. And so there is no end to life, to intellect, and the perfection of humanity. Its progress is everlasting.

Konstantin Tsiolkovsky

Konstantin Tsiolkovsky, who worked out the mathematics of rocketry and spaceflight, knew that chemical rockets by themselves would never become a practical, affordable means of space transportation that would allow mankind to settle space. It wasn't until after he had conceived the idea of the Earth surface to geostationary orbit tower as a way of making travel into space affordable, that he made this statement. The inspiration for the orbital tower came as a result of his visit to the Eiffel Tower in 1895. While he was aware that such an orbital tower could not be built with the technology of his time, just the fact that a theoretical solution to the rocket problem existed convinced him that someday a workable solution would be found.

Now, over 200 years after Reverend Malthus, the question of mankind escaping the limits to growth has yet to be resolved. Will we become a spacefaring civilization and move into a universe of unlimited possibilities where there is no limit to what we might become, or will we stay home and live with the limits to growth and all the limitations that will place on us?

1898 Guglielmo Marconi successfully demonstrated the utility of wireless radio communications between ships at sea and between ships and shore.

1900 Total human population is estimated to be 1.6 billion people.

1900 First flight of a Zeppelin airship.

1903 The Wright brothers build and fly the first successful heavier-than-air powered aircraft.

1905 Albert Einstein publishes the theory of special relativity.

1914 The beginning of World War I.

1916 Albert Einstein publishes the theory of general relativity.

1917 The Russian revolution brings down the Tsar and Russia becomes the world's first communist state.

1918 The end of World War I. Total death toll for the war is estimated at 17 million people.

1924 First crossing of the Atlantic Ocean by a Zeppelin airship.

1924 Following the death of Vladimir Lenin, Joseph Stalin becomes the unchallenged leader of the Soviet Union.

1925 Total human population is estimated to be 2.0 billion people.

1925-1926 Adolf Hitler publishes *Mein Kampf* (My Struggle), a book about his vision and plans for transforming German society into one based on the Aryan race being supreme over all others.

My son, if sinners entice you, do not consent.
Keep your feet from their path; for their feet run to evil,
and they hasten to thieve and shed blood.
Yet for all their cunning and self-deluded wisdom,
born of anger, hate, and unforgiven pain,
they lie in wait only for their own blood; they ambush only themselves.
Such is the end of all who go after ill-gotten gain;
it takes away the life of those who get it.

The Book of Proverbs, Chapter 1, versus 10-19

1927 Charles Lindbergh makes the first non-stop flight from New York to Paris.

1929 The *Graf Zeppelin* is the first airship to circumnavigate the globe.

1932-1933 Joseph Stalin orders the forced starvation of the Ukrainian people as punishment for their desire for national recognition and independence; an order that leads to the death of at least 6 million people.

One death is a tragedy; one million is a statistic. *Joseph Stalin*

1933 Adolf Hitler is appointed Chancellor of Germany.

1933-1945 German authorities, under the leadership of Adolf Hitler, establish a system of camps for confining political and ideological opponents as well as the "lesser races." The total number of people killed in these camps is estimated to be in the neighborhood of 17 million people by the time World War II is over.

Absolute power corrupts absolutely. *John Emerich Edward Dalberg Acton*

1934-1937 Joseph Stalin launches a massive purge, now known as the Great Purge, or the Great Terror, against internal enemies of the State that kills approximately 1 million people.

By the time of his death in 1953, Joseph Stalin is believed to be responsible for the death of somewhere between 30 to 40 million people, not counting the estimated 20 million Soviet military deaths in World War II.

Death is the solution to all problems. No man – no problem. *Joseph Stalin*

1936 The airship *Hindenburg* crashes on landing at Lakehurst, New Jersey. It is the end of the airship era.

1939 Germany invades Poland. It is the beginning of World War II.

1944 The V-2 rocket is the first rocket to reach space.

1945 The detonation of two atomic bombs over Japan brings about the end of World War II. Total death toll for the war is estimated at 60 million people.

1948 Bell Laboratories invents the transistor.

1949 Under the leadership of Mao Zedong, China becomes a communist state.

1950 Total human population is estimated to be 2.5 billion people.

1952 First commercial jet airliner enters service.

1958 Texas Instruments and Fairchild Semiconductors invent the integrated circuit.

1958-1962 A massive famine in China, due to a combination of flood, drought, and the communist policies of Mao Zedong, kills at least 45 million people.

1960 Oral birth control pills are approved for use in the United States. The fertility rate in the US at this time is approximately 3.6 births per woman.

1961 Yuri Gagarin orbits the Earth.

1963 The X-15 rocket plane flies into space.

U.S. Air Force photo

1964 Herbert Marcus, the man considered by many to be the father of the "New Left" in the United States, publishes *One-Dimensional Man*, a Marxist philosophy that inspired many radicals and political activists in the 1960s and 1970s. He is also considered to be the source of the Political Correctness movement in America. A movement that critics say is designed to inhibit freedom of speech, freedom of thought, and promote obedience and conformity.

History has shown that progress comes from differentiation, not uniformity.

Louis D. Brandeis

1966-1976 Mao Zedong initiates the Chinese Cultural Revolution; a communist-led purge of capitalist and traditional cultural elements within the country.

The sign of a true revolutionary is his desire to kill. *Mao Zedong*

It is estimated that up to 30 million people died as a result of this purge.
By the time of his death in 1976, it is estimated that Mao Zedong was responsible for the death of 75 million people.

Evil exists; and with darkness, sloth, laziness, divisiveness, anger, hate, fear, intolerance, ignorance, and mediocrity, it strives to take over the Earth.

Author

1968 Peter Glaser introduces the concept of satellite solar power stations for beaming clean, unlimited, low-cost power down to the Earth.

1968 Apollo 8 is the first manned spacecraft to leave Earth orbit and orbit the Moon. It was during this flight, while in orbit around the Moon, that a photograph of the Earth rising over the Moon was taken. This photograph, named *Earthrise*, has been described as the picture that started the environmental movement.

Earthrise *NASA photo*

1969 Gerard O'Neill proposes the building of Space Colonies as a method of space colonization and settlement.

1969 Apollo 11, Neil Armstrong and Buzz Aldrin land on the Moon.

NASA photo

We have taken to the Moon the wealth of this nation,
the vision of its political leaders,
the intelligence of its scientists,
the dedication of its engineers,
the careful craftsmanship of its workers,
and the enthusiastic support of its people.
We have brought back rocks, and I think it is a fair trade . . .
Man has always gone where he has been able to go. It is that simple.
He will continue pushing back his frontier,
no matter how far it may carry him from his homeland.

Michael Collins, Astronaut, Apollo 11

1971 The Apollo Program is running at full speed. The mood in America regarding manned spaceflight couldn't be better. We had succeeded, we had beaten the Russians to the Moon, and we were opening up the space frontier for all mankind. Most adults were of the mindset that while they would not be able to go themselves, their children would. They were creating a new world full of hope, promise, and unlimited possibilities for the next generation. It was a wonderful period in time.

It was during the preparations for one of the Apollo Moon flights that a newsperson made the announcement that each Moon flight cost in excess of $300 million. It was an incredible amount of money at the time and a stunned shock slowly settled over the country, especially among those who were avid supporters of the space program or who had dreams of going themselves. As the shock wore off, a new consensus began to take shape. It was that we had been living in a wonderful dream, a dream because at that price there was no way that space would become a frontier that was open to anyone with the desire to go. It was the beginning of the end for the Apollo Moon Program.

1971 Intel invents the first microprocessor.

1972 Apollo 17, the last landing on the Moon of the Apollo program.

1972 The fertility rate in the United States drops to 2.0 births per woman. The fertility rate in the US has remained between 1.7 and 2.1 ever since.

1975 Total human population is estimated to be 4.0 billion people.

1975-1979 A communist group known as the Khmer Rouge takes control of Cambodia and begins a purge that results in the death of 3 million people.

1977 The Personal Computer revolution begins with the introduction of the Apple II computer.

1979 *The Fountains of Paradise*, a book by Arthur C. Clarke, describes the building of an Earth surface to geostationary orbit Space Elevator that makes spaceflight so affordable that it opens up space and the solar system for settlement and development. It is this book that brings the idea of the Space Elevator to public awareness.

The Earth surface to geostationary orbit Space Elevator is based on the Earth surface to geostationary orbit tower that was first conceived by Konstantin Tsiolkovsky in 1895. The difference between them is simple; Tsiolkovsky's Tower is a compression structure that rests on the surface of the Earth like a really tall skyscraper. The Space Elevator is a tension structure that hangs down from a space station in geostationary orbit all the way to the surface of the Earth, like a rope hanging down from a tree house in a really tall tree. Unfortunately, and like Tsiolkovsky's Tower, the Space Elevator cannot be built with existing materials and technology.

1981 First flight of the Space Shuttle.

Space Shuttle Columbia *NASA photo*

1990 A non-rotating Skyhook, in combination with a reusable suborbital launch vehicle is proposed as a cost competitive alternative to the Earth surface to geostationary orbit Space Elevator that can be built with existing materials and technology.

1991 The World Wide Web is introduced.

1991 The Soviet Union collapses, and along with it Soviet-style communism.

When the history of communism, socialism, fascism, or any other type of government or social organization that demands racial, religious, or political conformity is examined in the cold light of reason, it is difficult not to be horrified by the ever growing list of purges, prison camps, death camps, mass murders, mass starvations, ethnic cleansings, public murders and beheadings, that have been initiated, supported, justified, and maintained by their proponents. Yet in spite of the ready availability of knowledge and information regarding this history, there continues to be large numbers of individuals who believe that these types of government and social organizations are the preferred way of life for mankind.

Many of these proponents justify their continued support based on their belief in the economic theories of Karl Marx; theories that lead to the economic collapse of the Soviet Union and which were abandoned by China when it returned to capitalism after the death of Mao Zedong. Another aspect of these types of government that does not get discussed very often is how the demand for conformity stifles productivity, innovation, and new ideas: a process that over time inevitably leads to the collapse of that social group, no matter how wealthy or powerful they might be when they start down this path.

1991 Carbon nanotubes are brought to the awareness of the scientific community.

1992-1995 Serbians, under the leadership of Slobodan Milosevic, a former communist, kill at least 100,000 people in the name of "ethnic cleansing", in former Yugoslavia.

1993 The Hubble Space Telescope becomes operational in Earth orbit.

1995 The Hubble Space Telescope takes a picture that has been described as "A myriad of galaxies back to the beginning of time." It is the first photograph that gives a view all the way to the visible horizon of the universe. *See page 145.*

1998 Fibers made of carbon nanotubes are proposed as a possible material that might make an Earth surface to geostationary orbit Space Elevator possible. While no carbon nanotube fibers have been made as of this date, proponents say that such fibers are only a few years away at most.

2000 Total human population is estimated to be 6.0 billion people.

2001 Dennis Tito becomes the first private individual to fly in space.

2004 SpaceShip One, an air-launched suborbital rocket plane, makes the first private manned flight into space.

D Ramey Logan

2011 Total human population is estimated to be 7.0 billion people.

2011 Last flight of the Space Shuttle.

NASA photo

2012 The Dragon spacecraft becomes the first commercial spacecraft to rendezvous and dock with the International Space Station.

NASA photo

2014 Small quantities of carbon nanotube fibers are finally made and tested. The strength of these fibers is only a tiny fraction of what is required to build a Space Elevator. They are not even as strong as existing graphite fibers. Proponents of the Space Elevator say they are 20 years away from building an Earth surface to geostationary orbit Space Elevator.

2015 First successful vertical landing of a reusable first stage booster for an orbital launch vehicle.

Chapter 3: Today

Without a clear vision of the future, we flounder in the present.

Dr. Sally Ride

Today, we stand on the brink of a new era of exploration and expansion into the solar system, and only two significant questions remain to be answered. Do we have the will to do it, and how can it be made affordable?

Shall we maintain the current status quo of space travel, so expensive that only a select few may go, or shall we look for a way to make it affordable to anyone who has the dream and the desire to go?

Will we choose to build a new "transcontinental railroad" to the edge of space, where we can build the new "clipper ships" to travel the solar system, or will we stay home?

Do we have the courage and the vision to create a new future where the opportunities for growth and what we might become are absolutely unlimited, or will we stay home living an increasingly restricted life where our choices are constrained by an ever growing population?

NASA artwork

Chapter 4: The Problem with Rockets

Every problem is an opportunity. *William Sahlman*

The three most important issues that need to be addressed if a transportation system is to be commercially affordable on a mass-market level are:

- reusability;
- a low propellant fraction; and,
- a high payload fraction.

It needs to be reusable so that the cost to build can be distributed over a large number of journeys. It also needs to be able to make the journey with a minimum amount of fuel or propellant so as to keep the operating costs as low as possible. Finally, it needs to be able to carry enough cargo that the cost of the journey can be distributed over the largest possible base. The final goal of all this is to reduce the cost of transportation to a small fraction of the value of the cargo. Good examples of transportation systems that do this are trains, planes, automobiles, trucks, and ships: expendable launch vehicles are not.

Reusability There are no reusable launch vehicles for going to Earth orbit. With the exception of the Space Shuttle, which was only partially reusable, they are all large, multi-stage, multi-million-dollar use once throw-away vehicles. This is due to the high propellant fraction required for going to orbit. In order to make a launch vehicle reusable, or to make the stages of a launch vehicle reusable, it is necessary to add additional weight in the form of parachutes or wings or additional control systems, possibly additional propellant for landing, and maybe some type of landing gear. If the part that is being made reusable experiences reentry temperatures, then it will also need some sort of heat shield. All of this adds weight, which takes away from the weight of the payload, and adds to the cost of the launch vehicle. If the weight increase for making the launch vehicle reusable exceeds the size of the payload it carries, then making the launch vehicle reusable has no purpose as it has no payload. This is the reason why all the previous attempts to build a 100% reusable orbital launch vehicle have failed, the weight increase for making them reusable exceeded the size of their payload.

It is a problem that exists even when the reusable launch vehicle has some remaining payload capacity. For example, assume that by making a launch vehicle reusable that it is possible to reduce the cost of flying it by 75%. If the additional weight of making the launch vehicle reusable reduces the size of the payload by only 50%, the end result is the cost per kilogram to orbit will go down. On the other hand, if the size of the payload is reduced by more than 75%, the cost per kilogram to orbit will go up.

As a result, the main problem with designing and building a reusable launch vehicle is designing one that can still carry a large enough payload to orbit so that it saves money. So far none of the attempts to design and build such a vehicle have been able to do this. All of them have had real or projected costs that were significantly higher than existing expendable launch vehicles. This is why there are no reusable launch vehicles.

A low propellant fraction The propellant fraction of a chemical rocket powered launch vehicle is determined by the amount of speed it needs to achieve to reach orbit and the propulsive efficiency of the propellant combination that it is using. In spite of the large differences in propulsive efficiency of the various propellant combinations, the propellant fraction of any past or currently existing ground-launched orbital launch vehicle is somewhere between 85% and 92%. This means that for every 100 kilograms of takeoff weight, somewhere between 85 and 92 kilograms of that are propellant, and 8 to 15 kilograms are the launch vehicle and payload. This is entirely due to the performance limitations of chemical rocket motors and the speed required for orbit. Current chemical rocket motors are already very close to achieving their theoretical maximum performance. As a result, there is no possibility of a significant reduction in propellant fraction in a ground-launched launch vehicle that uses chemical powered rocket motors as there is so little room left for improvement in the rocket motors. The only other ways to reduce the propellant fraction is to reduce the speed required for orbit or use a different type of propulsion system.

A high payload fraction Back in the beginning of the space age, the payload was the satellite or spacecraft that sat on top of the launch vehicle that was delivered to low Earth orbit. This type of payload is called the "gross payload to low Earth orbit", or "gross payload", in this book. In the case of manned spaceflight during this period, the gross payload was the spacecraft with the crew, their supplies, and any equipment they needed. Since the destination was low Earth orbit and the spacecraft served as the workspace and habitation module, it all counted as useful payload.

Now there are space stations in Earth orbit and the spacecraft serves as a delivery vehicle for carrying the crew, their supplies, and any equipment that is needed to the space station. Now it is no longer appropriate to count the spacecraft as useful payload. The useful payload is the crew, their supplies, and the equipment.

The useful payload delivered to a space station is always significantly less than the gross payload delivered to low Earth orbit by the launch vehicle. This might sound like a subtle distinction, but it makes a big difference when calculating the cost of going to a space station. The true cost of going to a space station is the total cost of the flight divided by the amount of useful payload that is delivered. This is what determines the true cost per kilogram for the journey.

Some examples of this from historic launch vehicles are:

Titan 2 GLV

Propellant fraction	91.5%
Empty weight fraction	6.1%
Gross payload fraction	2.4%

Saturn 1B

Propellant fraction	85.3%
Empty weight fraction	11.9%
Gross payload fraction	2.8%

If it is assumed that the useful payload delivered to a space station is one-third of the gross payload weight delivered to low Earth orbit, then the useful payload fraction for both of these launch vehicles is less than 1% of the take-off weight.

That is a very low useful payload fraction. By comparison, a typical automobile has a useful payload fraction of 20%, and an 18-wheel truck has a useful payload fraction of over 50%. Trains and ships are even higher. Commercial airliners operating at max take-off weight with a max fuel load have a useful payload fraction of between 8% and 15% depending on the aircraft.

Unfortunately, there is no way to significantly improve the useful payload fraction of current launch vehicles without noticeably reducing either the propellant fraction or the empty weight fraction. Since current launch vehicles already have the lowest empty weight fractions of any flight vehicles that have ever been built, there is not a lot of room left for improvement. All of these issues are why spaceflight based on the von Braun vision of large ground-launched multi-stage rockets is so expensive and why all those wonderful dreams of space colonies, satellite solar power stations, asteroid mining, and cities on the Moon and Mars are still only dreams.

Note All vehicles can be broken down into three major components: the weight of the usable propellant; the weight of the cargo or payload; and, the operating empty weight of the vehicle. It is the total of the three that makes up the weight of the vehicle at the start of its journey.
Payload fraction is the weight of the cargo or payload divided by the total weight of the vehicle.
Propellant fraction is the weight of the usable propellant divided by the total weight of the vehicle.
Empty weight fraction is the operating empty weight of the vehicle divided by the total weight of the vehicle.
The total of these three fractions must always equal 1. What that means is that if you add to one of them, you must take away from one or both of the others.

Chapter 5: A Shift in Thinking

When what you are doing doesn't work and you have exhausted all possible avenues of improvement, it is time to start thinking outside the box.

In the 1950s and 1960s a popular high school science experiment involved placing a rat in a maze and measuring how long it took the rat to find the piece of cheese that was hidden in the maze. Most students viewed this experiment as an intelligence test for rats. They would time their rat as it ran through the maze again and again and then plot the results as a way of measuring how quickly the rat was learning. They would then compare the learning curve of their rat with all the other student's rats and whoever had the smartest rat was the winner.

One day, one of the rats climbed out of the maze and walked across the top of the walls to the cheese in record time. This is an example of what most people call "out of the box thinking."

The owner of this rat thought this was proof that her rat was the smartest of them all. All the other students called it cheating. In order to keep the peace, the teacher put a piece of clear plastic on top of the maze so that it couldn't happen again. This is an example of how most people respond to "out of the box thinking."

A second experiment using the same rats and maze occurred when one of the students decided to move the piece of cheese to another part of the maze. By the time this change was made, the rats had already learned where the piece of cheese was so that little learning was going on anymore. The time measurements for finding the cheese had become more of a speed test to see which rat could run the fastest. Once the cheese was moved everything changed. As might be expected, the rats would race to where they had last found the cheese, look around in confusion for the cheese and then wait. Sometimes they would go back to the beginning of the maze and then run back to where the cheese had been. It was like they were checking to see if they had made a mistake, but then they would settle down and wait. They did this for days, returning to the previous location for the cheese and then waiting. Finally, one of the rats started searching the maze again and eventually found the new location for the cheese. In all cases, the rats took far longer to start searching for the relocated piece of cheese and far longer to actually find it then they had the first time at the beginning of the experiment. Maybe this is why the space programs of the world are having such a hard time changing from the von Braun vision for space exploration to one that will actually make spaceflight mass-market affordable.

Through the clouds, headed for orbit. NASA photo

Chapter 6: Learning from the Past to See the Future

Sometimes, when trying to stimulate "out of the box thinking", i.e., new ideas and new solutions to a problem, it pays to look at how similar problems were solved in the past. One possible problem from the past that applies to our current situation has to do with steamships.

Back in the beginning of the steamship age in the early 1800s, the steam engines of the day burned so much coal that the ships were limited in how far they could travel and still have enough room left over to carry a worthwhile cargo. They solved this problem by breaking up the longer shipping routes into shorter segments with strategically placed coaling stations where the ships could refuel. This allowed the early steamships to travel the globe while carrying less fuel (a reduced propellant fraction) and more cargo (an increased payload fraction). This reduced the cost of hauling the freight for those who had goods to ship and allowed the shipping companies to operate at a higher profit margin. It was a win-win solution for everyone.

So how does this apply to launch vehicles going from the surface of the Earth all the way to orbit when building a refueling station halfway up is not an option?

Lunar Module "Spider" in orbit around the Earth, Apollo 9, NASA

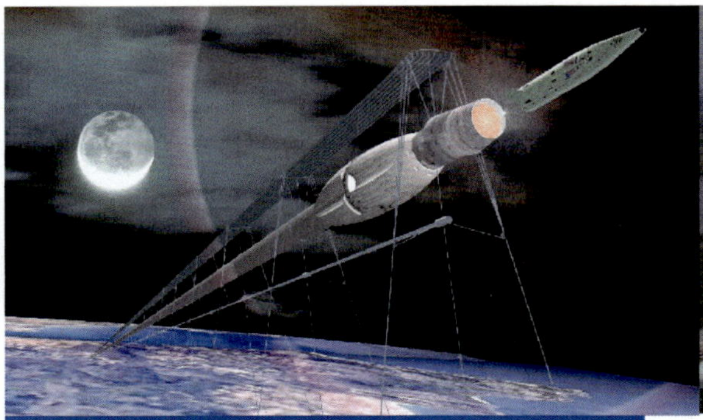

NASA photos and artwork

Chapter 7: What is a Rocket?

More accurately, what is the best way to think about or envision, a rocket that takes off from the surface of the Earth and flies into orbit at a cost that just about anyone can afford? Is it a large, super-fast airplane along the lines of the supersonic Concorde that flies into space? Is it a monster-sized single stage steel hulled rocket ship full of sound and controlled fury that ponderously roars its way into space as in a science fiction movie? Is it like one of the small shuttlecraft shown in "Star Trek": more of a space taxicab than a rocket? Is it a giant lighter-than-air blimp that floats to the top of the atmosphere where it then fires a small plasma thruster to lift it the rest of the way into space? Is it a 1,000-kilometer long tube that uses air pressure to accelerate a launch vehicle to orbital speed, similar to blowing on a soda straw to launch a small wad of paper across a room? Is it an elevator that climbs up a 100,000-kilometer long cable that goes from the surface of the Earth to geostationary orbit and beyond? What exactly is it? What is the vision that will lead us to an affordable to build, affordable to use, safe, reliable, launch system that we can build today?

When you look at the launch vehicles we have today, what we have are dragsters. Dragsters with really large motors and really large propellant tanks, where the propellant tanks are dropped off and thrown away when they are emptied, and the dragster keeps going, peddle to the metal, faster and faster, until it finally empties the last propellant tank. If we have planned everything correctly, the final speed of the dragster will be fast enough to stay in orbit. If not, it will fall back to Earth. They are awe-inspiring to watch. They put on an incredible show full of thunder, light, smoke, and flames. They roar so loud they make your teeth rattle and the Earth shake. They are thunder chariots that roar across the sky like an ancient Greek God. Unfortunately, as currently used, they will never be a mass-market affordable transportation system.

So how do we go from a thundering dragster that sheds propellant tanks as it roars across the sky, to something that would operate like a sleek, sedate, affordable steamship that safely goes about its business day after day?

We start by thinking of speed as an analog to distance. The steamship moves at a slow steady speed, slowly and continuously burning fuel such that over time it goes all the way around the world. The launch vehicle burns its propellant fast and furiously and then coasts all the way around the world. Assume that both use the same amount of fuel/propellant to get around the world. The rocket has to carry all of its propellant and has very little room left for cargo while the steamship has refueling stations all around the world, which leaves it with lots of room for cargo.

Notice that the common denominator between these two is fuel/propellant: fuel/propellant that can be used either for distance or for speed. When it is used for distance, it is possible to break the journey into smaller segments. When it is used for speed it is possible to reduce the required speed of the launch vehicle by using either air-launch or a catapult to boost the launch vehicle at the beginning of the journey, and by using a Skyhook to catch the launch vehicle at the end of the journey. Both of these reduce the speed required to reach orbit and as a result reduce the amount of propellant the launch vehicle needs to carry. This allows the size of the payload to be increased. Once the speed required is reduced enough, some of the increased payload capability can be used to make the launch vehicle reusable. When that happens, safe, reliable, mass-market affordable Earth to orbit transportation will become a reality.

Saturn V Moon Rocket preparing for launch, NASA photo

Chapter 8: Cheating

How does a rocket scientist think outside the box? How does a rocket scientist successfully cheat the laws of physics and make spaceflight mass-market affordable?

The answer is simple: use a different vision that reduces the required speed for orbit of the launch vehicle, limit yourself to concepts that can be built with existing technology, and think like a businessman.

Some people have asked, "Why, think like a businessman?"

Again, the answer is simple: if you want lower costs, look to free enterprise. America's strength has always come from free enterprise and the market competition that goes along with that. It has never come from government programs.

When President Eisenhower created NASA in response to the Soviet Union's launch of Sputnik, he did so with reluctance. Eisenhower believed that a space race with the Soviets would force us away from our core values of individualism, innovation, free trade, competition, democracy, and require that we create a large-scale institutionalized "technocratic nightmare" run by a "scientific-technological elite." Looking back over the last 50 years, if you compare our rate of progress in the field of launch technology with just about any other technological field such as aviation, electronics, software, computers, or the Internet, it would appear that President Eisenhower's concerns were valid.

Some basic rules for thinking like a businessman.

- Keep start-up costs to a minimum. Amortization of high start-up costs can easily overwhelm the cost savings of a new technology.
- Keep technical risk to a minimum. Higher technical risk inevitably leads to higher start-up costs.
- Starting a new business is risky, and a new spaceflight company using new technology does not have a proven flight record, a proven management team, or a proven customer base. Therefore, a realistic business plan based on the existing market that identifies real customers is essential. "Build it and they will come", rarely works with business endeavors that require a large initial investment.
- Start small and grow into long-term goals if at all possible. This keeps the start-up costs to a minimum and allows the business to prove itself to prospective customers over time.
- All of these rules should be considered when evaluating a new vision/idea/technology.

Different Visions There are many. The following is a partial list.

Ground accelerators These are devices that add velocity to the launch vehicle using ground-supplied power. This reduces the amount of propellant the launch vehicle needs to carry, increases the amount of payload it can carry, and adds to the possibility of making the launch vehicle reusable. Examples of ground accelerators are: aircraft catapults; rocket sleds; launch towers; mass drivers; and, launch loops.

There are proposed ground accelerators that would accelerate a launch vehicle all the way to orbital speed and beyond. These would require large technology development programs of uncertain cost, duration, and success.

Most proposed ground accelerators only accelerate the launch vehicle to somewhere between Mach 1 and Mach 3. A Mach 1 ground accelerator would not require any new technology development plus it would have the lowest start-up cost. A Mach 2 or Mach 3 ground accelerator would have a longer more expensive track and would require some development work due to the higher aerodynamic forces acting on the launch vehicle. As a result, the Mach 2 or Mach 3 ground accelerator would require a larger initial investment but would add more velocity to the launch vehicle, thereby making the launch vehicle smaller and more cost-effective.

*See **Chapter 10** for more information on ground accelerators.*

Air-launch to orbit This is when a launch vehicle is carried to high altitude by a conventional aircraft where the launch then occurs. Like a ground accelerator, this reduces the amount of propellant the launch vehicle needs to carry, increases the amount of payload it can carry, and adds to the possibility of making the launch vehicle reusable. Air-launch has been done at both subsonic, and supersonic speeds. Examples of subsonic air-launch systems are all the X-planes of the late 1940s through the 1960s; the best known being the B-52 launched X-15 rocket plane. In more recent times there has also been SpaceShip One and SpaceShip Two, both of which were air-launched from the White Knight One and White Knight Two carrier aircraft. Also currently under development is Stratolaunch, a subsonic carrier aircraft that will become the world's largest aircraft when it flies.

The only publicly known supersonic air-launch was the Lockheed D-21 drone that was launched from the back of the SR-71. This program had three successful supersonic launches but lost both the carrier aircraft and the drone during the fourth. All subsequent launches of the D-21 were made using a subsonic carrier aircraft. Based on this, the development of safe and reliable supersonic air-launch for a reusable launch vehicle would most likely require a large development effort of uncertain cost and duration.

*See **Chapter 11** for more information on air-launch to orbit.*

Air-breathing and rocket motor combination propulsion systems This is a concept that works by increasing the overall propulsive efficiency of the launch vehicle in order to reduce the amount of propellant the vehicle needs to carry. There have been many different types of combination air-breathing and rocket motor propulsion systems proposed with a wide range of performance improvements, increases in propulsion system weight, development costs, and risk levels. Some examples of these are air-augmented rockets, rocket/ramjets, rocket/ramjet/scramjets, turbine/rocket/ramjet/scramjets, and so on.

Only a small number of point design tests for a few of these concepts have been built, tested and/or flown in sub-scale tests. No full-size, entire speed range systems have been built and flown. Some of these concepts could be made operational with minimal effort. Others would require larger programs of uncertain cost and duration.

*See **Chapter 12** for more information on combination air-breathing and rocket motor propulsion systems.*

Reusable first stage A reusable first stage with expendable upper stage launch vehicle is a partially reusable launch vehicle that seeks to maximize the cost savings that come with reusability while keeping the weight gain to a minimum. The maximum cost savings occurs when the reusable first stage is as large as it can be without needing a heat shield, and the expendable upper stage is as small as possible. Unfortunately, this drives the design to a high staging velocity, which leads to a long downrange landing distance for the reusable first stage. This can be a problem for over water launches that lack a dry land landing site. Over water launches would require either a barge for landing out at sea or they would need a lower staging velocity with a short enough downrange distance that the reusable first stage can fly back to the launch site. Used with either over land or over water launch sites, the reusable first stage concept has the potential to generate significant savings in overall launch costs as long as the weight gain is kept under control.

Another way to use the reusable first stage concept is to air-launch it and have the carrier aircraft position itself so that the reusable first stage can land where the carrier aircraft took off.

*See **Chapter 13** for more information on reusable first stage launch vehicles.*

Beamed power propulsion This concept works by beaming power to the launch vehicle from ground-based sources. This increases the performance of the propulsion system resulting in a reduced propellant fraction, an increased payload fraction, and allows the building of a reusable launch vehicle. Beamed power systems have been proposed using both laser beams and microwaves for power transmission. While successful short-range, low-speed sub-scale flight tests have occurred, there have been no flight tests of a truly representative system that would apply to an orbital launch vehicle. This concept would require a major development effort of uncertain cost, duration, and success.

Tall towers and spaceports that float on top of the atmosphere These proposed concepts place the launch facility either on top of a 20-mile tall tower or on a gigantic floating airship that is designed to float 20 miles up in the air. Some of them have ground accelerators at the top, and others have conventional airports at the top for reusable space planes that fly into orbit. All of these concepts are designed to reduce the cost of getting to orbit by reducing the amount of propellant that the launch vehicle needs to carry. They do this by being above 95% or more of the atmosphere. This significantly reduces the drag of the launch vehicle and thereby reduces the necessary propellant for reaching orbit. Launching from 20 miles up instead of from sea level also eliminates the propellant the launch vehicle would have used overcoming the force of gravity to get to that altitude.

The problem with these ideas is the size of the initial investment needed to build them and the lack of a large enough market to justify the investment.

Space Elevators This is the elevator into space concept that goes directly from the surface of the Earth to geostationary orbit altitude and beyond using an elevator. No rockets are needed. It is a wonderful idea; unfortunately, it cannot be built with existing technology.

Skyhooks There are two types of Skyhook, rotating, and non-rotating. These are the space elevator concepts that do not reach down to the surface of the Earth and that require a suborbital launch vehicle for getting to the lower end of the cable. Skyhooks work by reducing the velocity the launch vehicle needs to achieve in order to reach orbit. Skyhooks can be affordably built with existing technology and they can be tailored to fit the existing launch market.
*See **Chapters 14**, **15**, and **16** for more information on non-rotating Skyhooks.*

Combination systems These are launch systems that combine two or more of these concepts. One possible combination would be a subsonic air-launched reusable first stage with expendable upper stage launch vehicle for flying to orbit. Another possible combination system would be a Mach 1 ground accelerator launched winged reusable rocket/ramjet spaceplane for flying to the bottom of a non-rotating Skyhook.

Just about any combination is possible. The difficult part is finding the combination with the lowest possible launch cost, the smallest possible initial investment, and the lowest possible risk, that can be sized for the existing launch market, and has the potential to grow as the market grows.
*See **Chapter 17** for more information on combination systems.*

Chapter 9: Narrowing Down the Choices

The great enemy of truth is often not the lie – deliberate, contrived and dishonest – but the myth – persistent, persuasive and unrealistic. *John F. Kennedy*

A Space Elevator goes from the surface of the Earth to 100,000 km out into space. The bottom of the cable is fixed to the ground and the cable is designed to support an elevator as it travels from the surface of the Earth up to geostationary orbit altitude and beyond. A Space Elevator means no more throw-away rockets, and no more meteor-like reentries through the atmosphere when returning to Earth. With a Space Elevator, everything is reusable, it has a low propellant fraction, a high payload fraction, and it is powered by clean, cheap, electric power. With a Space Elevator, getting into space is as easy and safe as riding on an elevator.

It is a beautiful, elegant, lovely idea.

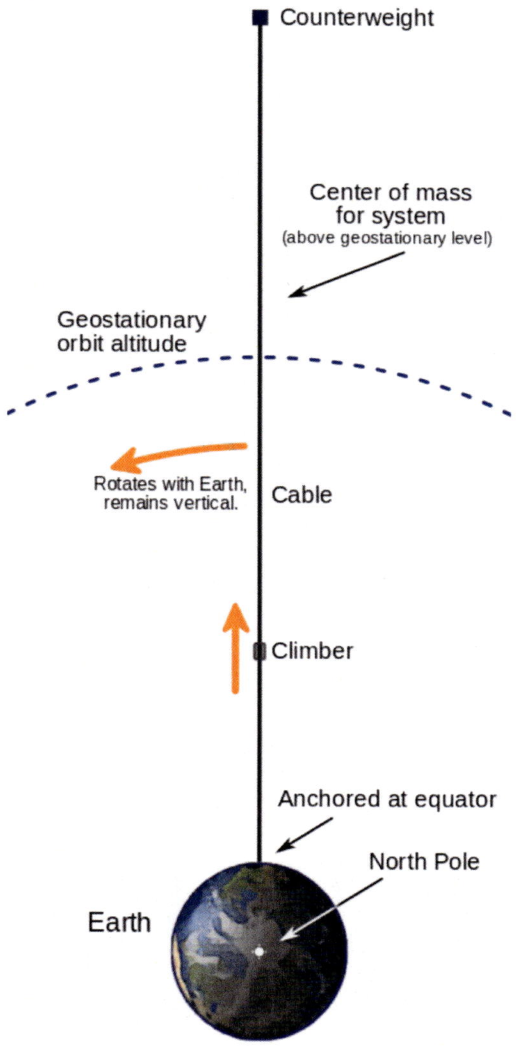

Unfortunately, it cannot be built with existing materials and technology.

This is the problem with many proposed futuristic launch concepts; they require major development programs of uncertain cost, duration, and success. A good example of this is the Space Elevator. After 14 years of development work on a fiber that people hoped would be strong enough to build it, the result is a product that is not as strong as existing graphite fibers. Even the proponents of this concept now say that a material strong enough to build the Space Elevator is 20 years or more in the future.

Many of the other visions that have been proposed for making spaceflight affordable share the same problems as the Space Elevator; they require new materials or technologies that currently do not exist. This makes them unrealistic contenders for building a mass-market affordable space transportation system that has what it takes to succeed as a business in today's market. So if we remove all the ideas and visions that require new materials, new technologies, or have national debt sized start-up costs, what is left?

- **Ground accelerators** with a speed between 600 MPH and 1,000 MPH. Faster speeds are possible but with a higher initial investment.
- **Air-launch to orbit**; specifically, subsonic air-launch to orbit.
- **Air-breathing and rocket motor combination propulsion systems**, that are limited to known technologies such as; turbines, afterburning turbines, ramjets, rockets, ducted rocket motors, ramjet-rockets, and ducted rocket/ramjet/rockets.
- **Reusable first stage** with expendable upper stage launch vehicles.
- **A non-rotating Skyhook**. This is the simplest, and least risky of the Space Elevator/Skyhook concepts that can be built with current technology. It is an idea that has been described as the equivalent of hanging a rope down from a treehouse so as to make lifting supplies and building materials to the treehouse easier.
- **Combination systems** made up from these concepts.

Elegance is simplification. As engineering designs evolve, they gain false sophistication – empty but seductive ingenuities. Ruthlessly, agonizingly, these must be stripped away.

Unknown

Chapter 10: Ground Accelerators

Ground accelerators are devices that add velocity to a launch vehicle using ground-supplied power. As uncommon and as little known as they are to most people, it is highly probable that the first rocket ever built was launched with a ground accelerator.

China It started with the invention of gunpowder in the 9th century in China. The first use of gunpowder was for making sparklers and flash paper for chasing away evil spirits. Both were popular items. Shortly thereafter, someone came up with the idea of making a hole in the end of a small joint of bamboo and packing it with gunpowder. The hole was then filled with a small fuse and the firecracker was born. This also appears to have been a popular item and many were made. Somewhere along the line, most likely in the rush to fill a purchase order, the hole in the end of the bamboo was made larger in order to make it easier to fill with gunpowder. This led to a lot of duds, instead of going "pop" they would skitter around on the ground going every which way. Occasionally, one of these duds would hit something on the ground and launch itself into the air where it would corkscrew around until all the gunpowder was burned. This led to the invention of two things. The first was the use of quality control inspectors at the firecracker factory for making sure the hole was the correct size. The second, and far more important invention, was made by the bright soul who decided to attach one of these defective firecrackers to an arrow and shoot it into the sky with his bow. Thus was born the world's first ground accelerator launched rocket, the fire arrow. You could even say this invention was the ever so humble beginning of the aerospace industry.

The Wright Brothers Ground accelerators have also played a role in more modern times. The Wright brother's first powered airplane, the Wright Flyer, was launched into the air using a ground accelerator. The engine for the Flyer barely had enough power for sustained level flight, let alone the power needed to accelerate up to flight speed. There was the additional problem of launching from soft sand. Everyone knows how difficult it is to walk in soft sand. Try rolling an airplane through it. That soft sand added tremendously to the power required for take-off, power that the Flyer did not have. The Wright brothers solved these problems by launching the Flyer with a track laid on top of the sand that also included a drop-weight powered catapult. It worked and the rest is history.

The V-1 and the V-2 In World War II the Germans invented the world's first cruise missile. They called it the V-1. It was a ground accelerator launched flying bomb. It has also been reported that the Germans launched a winged V-2 rocket with a ground accelerator from an underground tunnel in the closing days of World War II. They did this in order to extend the range of the V-2.

V-1 with ground accelerator at the Imperial War Museum, Duxford, UK

Today The United States NAVY uses ground accelerators to launch aircraft from the decks of aircraft carriers all around the world. Ground accelerators are also used to winch launch sailplanes into the sky. They have been included in science fiction novels and movies. One novel, *The Man Who Sold the Moon*, by Robert Heinlein, described a ground accelerator built on the side of Pike's Peak for launching rockets. It was published in 1949. Dr. Wernher von Braun, who was the technical consultant for the 1951 movie *When Worlds Collide*, also included a ground accelerator in that movie for launching the spaceship that carried a small group of people to another world.

NASA

Ground accelerators are known technology, they are affordable and they work. The rocket sled track at Holloman Air Force Base in New Mexico holds the world land speed record of 6,453 MPH for a tracked vehicle.

Types of ground accelerators

Horizontal and Mountain Slope Ground Accelerators Ground accelerators come in a number of forms. The most commonly used design is the horizontal track ground accelerator. The ones used by the NAVY on its aircraft carriers and the ones at Holloman Air Force Base all have horizontal tracks.

Most proposals for using a ground accelerator to boost a launch vehicle have horizontal tracks that are placed on the side of a mountain so they will have an upward slope.

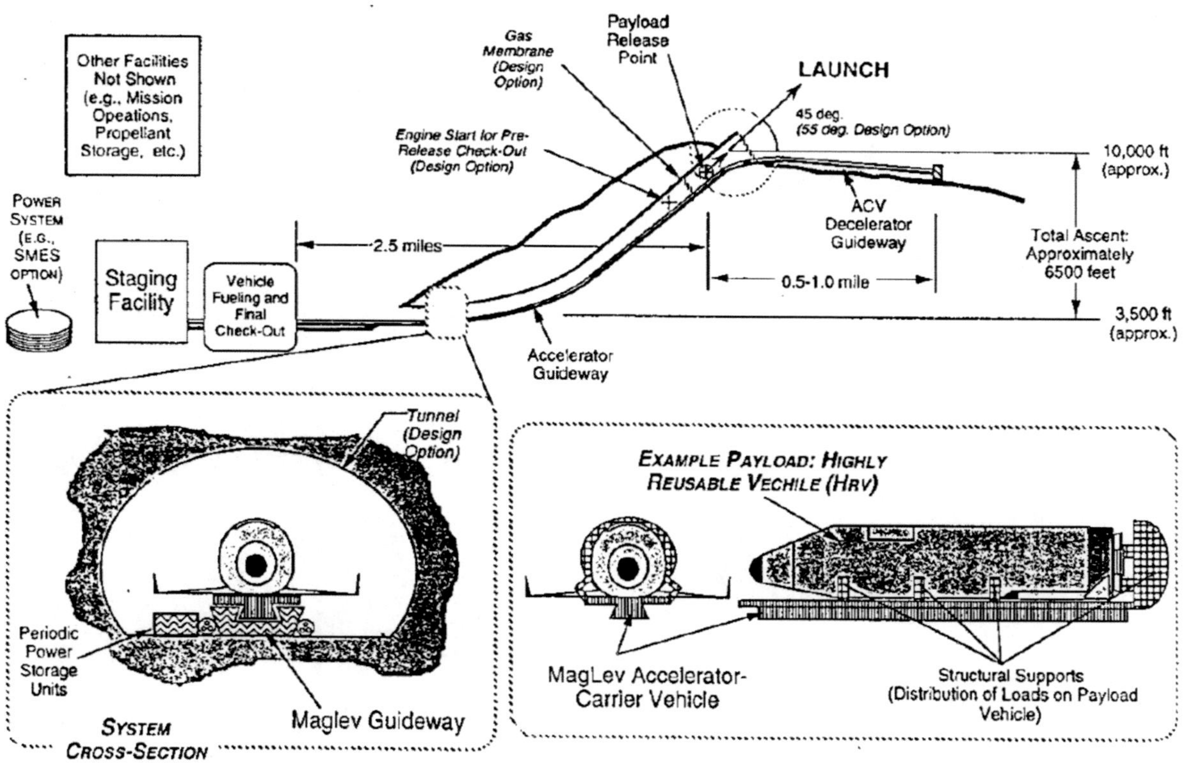

A Mountain Slope Ground Accelerator *for launching a Single Stage to Orbit launch vehicle.* ***From*** *the NASA study The MagLifter – An Advanced Concept Using Electromagnetic Propulsion in Reducing the Cost of Space Launch. See reference 17.*

Winch Launch Ground Accelerators The second most common form of the ground accelerator is the winch launch system that is used to launch sailplanes. This system is sometimes referred to as a trackless ground accelerator as the aircraft being launched rides on its wings as it is accelerated up to speed.

Vertical Accelerator Launch Tower The third form of a ground accelerator is called VALT, which stands for Vertical Accelerator Launch Tower. This is a horizontal track ground accelerator that has been placed on the side of a vertical tower. It launches the rocket straight up.
For more information on VALT see reference 18.

A small rocket leaving the upper end of a Vertical Accelerator Launch Tower, LEFT: prior to rocket motor ignition, RIGHT: after rocket motor ignition.

Advantages and Disadvantages of Each

There are advantages and disadvantages to all of them. The horizontal track ground accelerators are limited to what orbits they can reach due to their fixed orientation on the ground. The main disadvantage of this is that it limits the number of potential customers who will want to use it.

The second disadvantage of the horizontal track ground accelerator is the energy loss that occurs when the launch vehicle changes its flight path angle from the horizontal angle of the ground accelerator to one that allows it to climb out of the atmosphere and head for orbit. This is why all the proposed horizontal track ground accelerators for launching rockets are placed on a mountainside with an upward slope.

The third disadvantage of a horizontal ground accelerator is that the launch vehicle will require wings. Wings mean more drag, a slightly higher speed required for orbit, more empty weight and less payload. A mountain slope ground accelerator will be able to avoid this requirement as long as the upward slope of the track is steep enough.

The advantages of the trackless or winch launch ground accelerator as compared to the horizontal or mountain slope ground accelerator are; it is cheaper to build, and it can be designed as an all-orbit launch system.

It has two disadvantages; the launch vehicle will need wings just like the horizontal track system, and no one has ever built a winch launch system for aircraft this large or for going this fast before.

The Vertical Accelerator Launch Tower is also an all-orbit ground accelerator. In addition, this type of ground accelerator does not require the addition of wings to the launch vehicle, nor does it require as much land as the other two concepts. Therefore, it can be placed just about anywhere. The tower can be either free standing like the Eiffel Tower, or wire braced like an antenna tower. The tower can also serve as an observation tower and tourist attraction between launches. The main disadvantages of this concept are the cost of building the tower and its height. For example, a 600 MPH tower that accelerates the launch vehicle with 6 times the force of gravity will need to be 600 meters tall. That is twice the height of the Eiffel Tower.

Advantages of a Ground Accelerator Versus No Ground Accelerator

When reducing the speed required for orbit from the front end of the flight profile as a ground accelerator does, it saves more than just the speed it provides. It also saves the propellant the launch vehicle would have burned overcoming the force of gravity to get to that speed, the extra propellant used to overcome the aerodynamic drag, and the extra propellant burned due to the efficiency losses of the rocket motor from atmospheric pressure. An example of just how much propellant this is can be found on NASA's website in the article entitled *Space Shuttle Basics*. Using the numbers provided in that article, it turns out that the Space Shuttle burned 42% of its propellant just to reach 1,000 MPH. That is a lot of propellant. While a 600 MPH ground accelerator will not save that much, it does give an idea of just how large an impact the use of a ground accelerator can have.

When described in terms of velocity reduction, a 600 MPH mountain slope or VALT style ground accelerator located on top of a 6,100 meters (20,000 feet) tall mountain will reduce the required velocity of an orbital launch vehicle by approximately 900 meters per second. If the ground accelerator is located at 3,660 meters (12,000 feet) of elevation, the velocity reduction is approximately 800 meters per second.

International Space Station, NASA photo

Chapter 11: Subsonic Air-Launch to Orbit

Air-launch is another well-known technology that can significantly reduce the velocity required for orbit. Like the ground accelerator, it reduces the speed for orbit by more than just the speed it provides. It also saves the propellant the launch vehicle would have burned overcoming the force of gravity to get to that speed and altitude, the propellant used to overcome the aerodynamic drag while getting to that speed and altitude, and the extra propellant burned due to the efficiency losses of the rocket motor that come from operating in the atmosphere.

B-52 Carrier Aircraft with X-15 Rocket Plane under its wing, NASA photo

---

Its disadvantages are the need to add wings to the launch vehicle due to its horizontal launch attitude, and the cost to develop and build a large carrier aircraft when the launch vehicle is too heavy to be carried by an existing aircraft. A good example of just how big and expensive this can be is the Stratolaunch carrier aircraft that is currently being developed by Vulcan Aerospace.

The Stratolaunch carrier aircraft, artwork courtesy of Vulcan Inc.

When compared with either a horizontal track or mountain slope ground accelerator, air-launch has the advantage of being all-orbit capable. Air-launch also has the advantage of not subjecting the fully loaded launch vehicle to high acceleration loads like all of the ground accelerator concepts do.

There are other potential advantages to air-launch when compared with a ground accelerator, most of which depend on the type of launch vehicle being flown.

For example, when using a reusable first stage launch vehicle there is the issue of where the reusable first stage will land and how it will be returned to the launch site. If the launch vehicle is VALT launched from a coastal site on an over water flight, the reusable first stage will need to land either in the ocean, on a barge out at sea, or carry enough extra propellant so that it can fly back to the launch site – all of which will add to the cost of the system. If the reusable first stage launch vehicle is air-launched, the carrier aircraft can position itself so that the reusable first stage can land at the airport where the carrier aircraft took off. On the other hand, if the vehicle to be air-launched is a Single Stage To Orbit vehicle, or a Single Stage To Skyhook vehicle, the landing advantage will no longer apply.

Speed Required for Orbit

The speed required to reach low Earth orbit using a winged launch vehicle is approximately 9,500 meters per second (m/s). The total reduction in speed required for this orbit that is achieved from air-launching at 11,900 meters (39,000 feet) of altitude and at a speed of 575 MPH, is approximately 1,100 meters per second. That reduces the speed required for this orbit to approximately 8,400 meters per second.

A vertical take-off launch vehicle without wings or strap-on boosters needs to achieve a speed of approximately 9,100 meters per second in order to reach low Earth orbit. A 600 MPH ground accelerator reduces that speed by approximately 800 to 900 meters per second depending on how high the ground accelerator is above sea level. This reduces the speed required for low Earth orbit to approximately 8,300 meters per second.

Air-launch provides all-orbit capability from any airport as long as the carrier aircraft has the range to fly to the equator. An all-orbit ground accelerator that is located on the equator can do the same. If the ground accelerator is located north or south of the equator, the latitude of its location will limit the orbits it can reach.

Which is Better, Air-Launch or a Ground Accelerator?

The answer to that depends on a number of issues. Are they being compared as part of a suborbital launch system, as part of an orbital launch system, or as part of a combination launch system that includes a reusable first stage and a Skyhook? It will also depend on the long-range goals of the company and what part of the launch market they want to focus on. Politics and legal requirements are also involved.

Currently, both Vulcan Aerospace and Virgin Galactic are developing their own launch vehicles for launching satellites into orbit from their respective carrier aircraft and it is unlikely they would consider launching a competitor's launch vehicle with their carrier aircraft. To do so would mean selling fewer flights on their own launch vehicles, plus it would mean putting their carrier aircraft at risk carrying a launch vehicle they can never know as well as one that they were involved in both designing and developing.

The size of the carrier aircraft also puts a limit on the size of the launch vehicle, which places both an upper and lower limit on the size of the payload that the system can cost-effectively launch. This limits the number of potential customers and the number of flights per year for the launch system. This is important because a higher flight rate helps to reduce the cost of getting to orbit. In order to overcome this limitation, Vulcan Aerospace is planning to develop a number of different sized launch vehicles for Stratolaunch so they can cost-effectively launch a much wider range of payload sizes.

By comparison, a ground accelerator never leaves the ground so it is never at risk of total loss in the event of a launch vehicle accident like a carrier aircraft is. As a result, it makes much more sense to operate a ground accelerator more like a general service launch pad and design it for as many different types and sizes of launch vehicles as possible. This will allow the ground accelerator to operate at a much higher flight rate.

The final advantage of a ground accelerator over a carrier aircraft is the lower cost to develop.

The disadvantages of a ground accelerator have to do with its fixed location and the complications this creates when dealing with a combination launch system.

For example, when a ground accelerator launched reusable first stage launch vehicle for flying to the lower end of a Skyhook has a flight rate of multiple flights per month, it will be necessary to place the ground accelerator somewhere on the equator, at as high an elevation as possible, and for the Skyhook to be placed in an equatorial orbit. There will also be the issues of providing a downrange landing site for the reusable first stage, as well as providing whatever equipment and infrastructure are needed to get those stages back to the launch site. Setting up an operation like this in a foreign country brings up a number of issues regarding the political stability of the host country or countries, as well as legal issues concerning the export of launch vehicle technologies from the United States. These issues could represent a show-stopping disadvantage for an equatorial based ground accelerator launch system.

When an air-launched reusable first stage booster is used in combination with a Skyhook it is not necessary to place the Skyhook in an equatorial orbit in order to achieve a high flight rate, as the carrier aircraft will be able to position itself as needed in order to rendezvous with the Skyhook. Since the reusable first stage will require wings as a result of being air-launched, this will allow the reusable first stage to land horizontally at already existing airports. Having wings will also greatly increase the number of airports the booster will be able to reach after it has finished sending the upper stage and payload on its way to the Skyhook. This launch and landing site flexibility will also allow the launch company the option of operating from a non-equatorial country should that prove to be the better choice after all the legal and political issues are considered.

Another synergistic technology that should be considered in combination with air-launching a reusable first stage launch vehicle is the addition of some amount of air-breathing propulsion to the reusable first stage.

The Northern Lights as seen from space, NASA photo

Chapter 12: Air-breathing & Rocket Motor Propulsion Systems

A proposed air-breathing and rocket powered launch vehicle, NASA artwork

Internal combustion engines work by combining oxygen from the atmosphere with fuel. This air-fuel mixture is then burned to create heat and pressure, which is converted into motive power.

A jet engine also works by combining oxygen from the atmosphere with fuel. The air-fuel mixture is then burned to create heat and pressure, which is used to make thrust.

A rocket motor also works by combining oxygen with fuel. The oxygen and fuel mixture is then burned to create heat and pressure, which is used to make thrust.

The main difference between a rocket motor and the other two is that the rocket needs to carry its supply of oxygen along with it, while the internal combustion engine and the jet engine get their oxygen from the air.

The oxidizer to fuel ratio for a liquid oxygen, liquid hydrogen rocket motor is approximately 7 pounds of liquid oxygen for every pound of liquid hydrogen. That means that a liquid oxygen/liquid hydrogen powered launch vehicle with a 90% propellant fraction is 78% liquid oxygen by weight at take-off.

The idea behind using air-breathing propulsion for a launch vehicle is to get as much oxygen as possible from the air so as to minimize the amount of oxygen the launch vehicle has to carry to reach orbit. The lower the propellant fraction, the higher the payload fraction, and the lower the cost of getting to orbit.

Back in the late 1980s, there were a number of attempts to design and build a launch vehicle that incorporated air-breathing propulsion all the way to orbit. None of them were successful. All of them found that 100% air-breathing to orbit systems were beyond current technology. However, this does not mean that combination air-breathing and rocket motor systems that still carry some amount of oxygen will not work. As it turns out, there have been a number of studies and hardware experiments that show they do.

One concept called Argus, consisted of a 545 MPH ground accelerator launched winged reusable Single Stage To Orbit (SSTO) launch vehicle that was powered by a combination ramjet and ducted rocket motor propulsion system.

There was another concept called Northern Star, which was a Horizontal Take-Off Horizontal Landing (HTOHL) reusable first stage with expendable upper stage launch vehicle that was powered by a combination afterburning turbine, ducted rocket motor, ramjet, and scramjet propulsion system.

NASA's GTX was a third. GTX was a vertical take-off SSTO launch vehicle that used a combination ducted rocket, ramjet, scramjet, rocket motor propulsion system. There have been others and they all appeared to work.

Another type of air-breathing propulsion concept that could be used in a launch vehicle is the Liquid Air Cycle Engine (LACE). This type of propulsion system uses its onboard liquid hydrogen to liquefy the oxygen from the outside air and then combines that liquid oxygen with the liquid hydrogen in a rocket motor. Like all combination air-breathing/rocket motor propulsion systems, this concept only works while the launch vehicle is still in the atmosphere, after which it needs to switch over to using liquid oxygen from an onboard supply.

A variation on this concept is called SABRE, which stands for Synergistic Air-Breathing Rocket Engine. This system uses its supply of onboard liquid hydrogen to cool the compressed air from the outside air intake instead of liquefying it. This low-temperature compressed air is then fed into the rocket motor where it is combined with the liquid hydrogen and burned. This is the propulsion concept that is being developed for the horizontal take-off horizontal landing Single Stage To Orbit launch vehicle called Skylon.

The reduction in propellant fraction made possible by these propulsion concepts, their empty weight fractions, payload fractions, plus the level of risk and cost to develop, cover a wide range. There are also a number of variations on these concepts related to the types of propellants that are used, the transition velocities where they shift from one mode to another, and the other propulsion concepts that they could be combined with, which have not been investigated. So it is possible that the best answer is still unknown. It is also possible to combine a vehicle powered by one of these propulsion concepts with air-launch, a ground accelerator, and a Skyhook. All of these are possible additions that would add to the cost-effectiveness of these combination air-breathing and rocket motor propulsion systems.

Chapter 13: Reusable First Stage Launch Vehicles

The idea of making the stages of a multi-stage rocket reusable has been around almost as long as the idea of a multi-stage rocket. Dr. Wernher von Braun's multi-stage Ferry Rocket, as described in Collier's Magazine in the early 1950s, had both its first and second stages splashing down in the ocean where they were to be recovered for reuse. A real world example of this were the reusable solid rocket boosters from the Space Shuttle that splashed down in the ocean. Unfortunately, making a stage reusable makes it heavier, which reduces the amount of the payload the launch vehicle can carry. Reusability also has the additional costs of recovery and refurbishment. The problem with making both stages reusable is that the size of the payload goes down by a larger percentage than the reduction in cost to fly. The end result is that the cost per kilogram to orbit actually goes up instead of down.

The basic idea behind the reusable first stage with expendable upper stage concept is to minimize the weight increase that goes with reusability while keeping the majority of the cost savings. This works because the first stage does not have to deal with really high reentry temperatures like the upper stage does, thereby avoiding the significant weight penalty associated with adding a heat shield. Also, because the first stage is always larger than the upper stage, it allows for the largest and most expensive part of the launch vehicle to be recovered and reused. The end result is a partially reusable launch vehicle where the cost to fly goes down by a larger percentage than the payload size does. The end result is that the cost to orbit goes down instead of up.

There are three basic methods for recovering a reusable first stage:

1) put wings and landing gear on the first stage so that it can glide to a landing at an airport;

2) make the first stage into a vertical lander like Blue Origin and SpaceX are doing with their launch vehicles; and,

3) use Wernher von Braun's idea and have the first stage splash down at sea where it can be recovered by ship. This is probably the lowest development cost, lowest tech answer as far as flight hardware goes, but it does have the additional cost of keeping an ocean going tug and crew on station near the splashdown point when a launch is planned. It also has the problems of weather and sea conditions at the splashdown site, plus all the additional refurbishment costs that go with landing in salt water.

All of these ideas will work. The magic lies in determining which one reduces the cost per kilogram to orbit the most.

Cost Savings

When looking at the weight figures for historic launch vehicles, the first stage of a two-stage launch vehicle averages about 75% of the empty weight of the launch vehicle. If it is assumed that both stages have a similar cost per kilogram to build, then the first stage represents approximately 75% of the cost of the launch vehicle. If it is assumed that the reusable first stage has an initial flight life of 10 flights and that the refurbishment costs are 10% of the stage cost per flight, then the cost savings per flight should be somewhere around 60%. If there is no reduction in payload size, this 60% reduction in launch cost will also become a 60% reduction in cost per kilogram to orbit. If there is a 10% reduction in payload size due to a heavier first stage, the cost per kilogram will be reduced by 56%. Exactly how much the cost per kilogram to orbit will be reduced will be determined by both the reduction in launch costs and how much the size of the payload goes down. If it is assumed that the reusable stage has an unlimited flight life, no refurbishment costs, and no weight gain, the absolute best that can be achieved with this launch concept will be a 75% reduction in cost per kilogram. This is why combining the reusable first stage concept with a ground accelerator, air-launch, and/or a Skyhook becomes so important as all three of these additional concepts will add to the size of the payload. It is only in the combination of these ideas that spaceflight can be made affordable to everyone.

New Shepard: the world's first Vertical Take-Off Vertical Landing space vehicle. Photo courtesy of Blue Origin

Chapter 14: A Non-Rotating Skyhook

I cannot comprehend how any man can want anything but the truth.

Marcus Aurelius Antonius

Of the five items that made it to the list of visions to consider, the non-rotating Skyhook is probably the least known and understood of them all. So what is a non-rotating Skyhook? To start with, a non-rotating Skyhook is not a Space Elevator. While they are often thought to be the same thing, and they operate on similar principles, in truth they are very different.

The first and most important difference between a non-rotating Skyhook and an Earth surface to geostationary orbit Space Elevator is that the Skyhook can be built with existing materials and technology, while the Space Elevator cannot. The Space Elevator is a beautiful, elegant, lovely idea, but because the materials and technology to build it do not exist, it falls under the category of a beautiful dream. On the other hand, a Skyhook can be designed, engineered, and built, right now. There is no need to wait for new materials to be invented or new technologies to be created. There is no need to make educated guesses regarding what it will cost to build, how much it will weigh, or what it will cost to use. With the Skyhook, all of these details can be accurately calculated because all the materials and technology required to build it already exist.

The second major difference between a non-rotating Skyhook and a Space Elevator is that the Skyhook is much shorter and does not reach down to the surface of the Earth. The lower end of the non-rotating Skyhook is above the upper edge of the atmosphere and it requires a high-speed aircraft/suborbital launch vehicle to get to it. The advantage of the non-rotating Skyhook over existing launch vehicles is that the lower end of the Skyhook is moving at less than orbital velocity for its altitude. This means that a launch vehicle flying to the bottom of the Skyhook can carry a larger payload than it could carry to orbit on its own. This is similar to a ground accelerator that reduces the speed required for orbit by adding velocity to the launch vehicle at the beginning of its flight profile. In the case of the Skyhook, it reduces the speed required for orbit by providing a docking port that is moving at less than orbital velocity at the launch vehicle's destination altitude. Like the ground accelerator, this reduces the required propellant fraction of the launch vehicle, increases its payload fraction, and thereby reduces the cost per kilogram of getting to orbit.

This is what a Skyhook is. It works on the same principles that make an Earth surface to geostationary orbit Space Elevator work but it has been designed to work within the limits of currently existing technology. Put in the simplest possible terms, a non-rotating Skyhook is a partial Space Elevator that can be built today. It is that simple.

The Treehouse/Skyhook Analogy

Building a treehouse is a lot of work, as every piece of wood and all the tools to build it must be carried up the tree. This is especially true when it takes both hands and both feet to climb the tree. Another important issue is how high a person needs to climb to get to the treehouse. The higher up it is, the more work it takes to get there. So most people usually take a rope up with them early in the construction process so they can use the rope to lift all the wood and tools to the treehouse. Quite simply, it makes building a treehouse much easier.

This is what a non-rotating Skyhook is, it is the rope that hangs down from the treehouse that is used to haul things up and down. Only in this case the treehouse is a space station that orbits the Earth and the rope isn't long enough to reach all the way to the ground. Even if the rope did reach all the way to the ground, the bottom end of the rope would still be moving too fast to grab. So instead, the lower end of the rope is hung down to just above the upper edge of the atmosphere and a high-speed aircraft/suborbital spaceplane is used to fly to it. It still isn't easy to get to the lower end of the Skyhook, but it is much easier than going all the way to the space station without the Skyhook.

The Secret of the Skyhook

The secret of how the non-rotating Skyhook works has to do with how the speed of orbit goes down as altitude increases. The first step in taking advantage of this reduction in speed that comes with increased altitude is to place a space station in a higher slower orbit. Unfortunately, flying directly to this higher slower space station does no good as it takes more energy to boost the launch vehicle to the higher altitude than is saved by the slower speed of that orbit.

This is where the non-rotating Skyhook becomes important. The non-rotating Skyhook is a vertically oriented cable that is attached to the space station. Since the cable is moving with the orbital speed of the space station, the velocity of the downward hanging part of the cable is moving too slow to stay in orbit at the lower altitudes. As a result, the lower part of the cable wants to fall; only it can't because it is attached to the space station. So it hangs there, forever pointing straight down at the Earth. Now the lower end of the cable becomes the perfect docking port for a launch vehicle flying up from the Earth; a docking port that moves with the slower speed of a higher altitude orbit but at a lower altitude. The higher up the space station is, the longer the downward hanging cable can be, and the greater the reduction in speed at the lower end of the cable. The larger the speed reduction is, the greater the cost reduction becomes to fly there. When the cable is long enough, it then becomes possible to build a completely reusable single stage launch vehicle that operates like an airliner. This adds even more to the cost savings. This is called Single Stage To Skyhook flight.

Other Differences

As already mentioned, the Skyhook is much shorter than a Space Elevator. An Earth surface to geostationary orbit Space Elevator without an asteroid counterweight has an overall length of approximately 100,000 kilometers. A Space Elevator with an asteroid counterweight has an overall length of approximately 47,000 kilometers. Anything less than that for either one of them and they will not work. The Earth surface to geostationary orbit Space Elevator is an all or nothing concept as far as length is concerned. On the other hand, a non-rotating Skyhook can start out as short as 200 kilometers in overall length and it can be incrementally lengthened and strengthened over time to meet the needs of a growing market. Obviously, it is much easier and cheaper to start out building a 200-kilometer long cable than it is a 100,000-kilometer long cable. It is also important to keep in mind that every increase in the overall length of the Skyhook reduces the speed required for orbit: speed reductions that increase the payload fraction of the launch vehicle and further reduce the cost of getting to orbit.

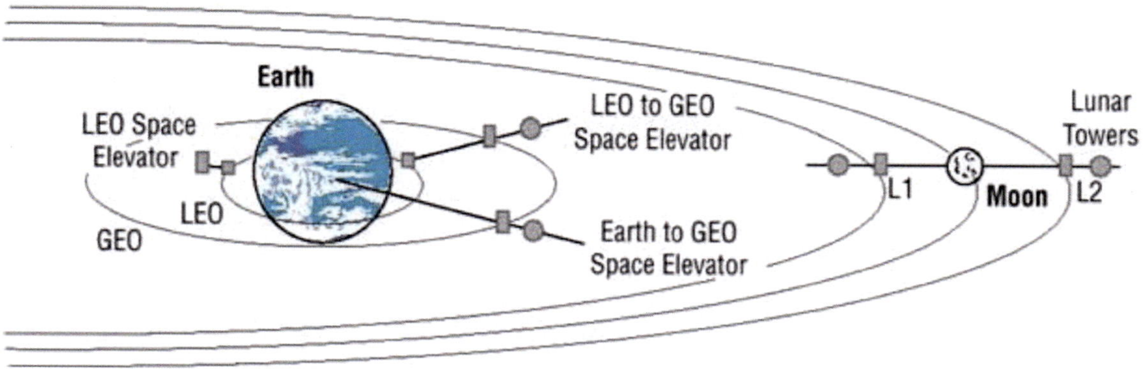

A visual comparison of the various Space Elevator and non-rotating Skyhook concepts that have been proposed (the non-rotating Skyhook is referred to as a LEO Space Elevator in this drawing), NASA artwork

---

Analysis has shown that the maximum cost-effective overall length for a non-rotating Skyhook based on currently existing graphite fibers is approximately 4,000 kilometers. That is 1/25th the length of a 100,000-kilometer long Space Elevator. Analysis has also shown that the reduced cost of building a 4,000-kilometer long non-rotating Skyhook as compared to a 100,000-kilometer long Space Elevator more than makes up for the additional cost of the suborbital launch vehicle that the Skyhook requires. As a result, a mature 4,000-kilometer long non-rotating Skyhook with reusable single stage suborbital launch vehicle is considered to be cost competitive with what is thought to be realistically achievable using a Space Elevator assuming a Space Elevator can ever be built.

Another difference between a non-rotating Skyhook and a Space Elevator is that the Space Elevator remains over the same spot on the Earth while the Skyhook circles the planet every few hours. This allows the Skyhook to serve as a terminal for launch vehicles arriving from just about anywhere on Earth.

The Upper End of a Non-rotating Skyhook

In the same way that the lower end of a Skyhook is moving at less than orbital velocity for its altitude, the upper end of the Skyhook is moving faster than orbital velocity for its altitude. This is due to the fact that the speed of orbit goes down as distance from the Earth goes up. Since the upward pointing cable is moving with the speed of the space station it is moving too fast to remain in orbit at the higher altitudes. This excess velocity makes the cable want to move away from the Earth and away from the space station and into a higher orbit. The end result is an upward pointing cable that points away from the Earth. The advantage of this is that a satellite or spacecraft that is released from the upper end of the Skyhook receives a velocity boost from the Skyhook without having to use any of its onboard propellant. The amount of velocity boost that the satellite or spacecraft receives depends on the length of the upper cable, the longer the cable, the greater the boost. In the case of a mature Skyhook, the upper end of the cable is designed to be moving at just short of escape velocity for its altitude. This means that a satellite or spacecraft departing from the upper end of the mature Skyhook can be placed in a geosynchronous transfer orbit, a free return orbit to the Moon, or sent on its way to a near-Earth asteroid, without the need for a large expensive expendable upper stage for boosting it to escape velocity. Elimination of the upper stage will also reduce the number of flights to the lower end of the Skyhook, which will further reduce the overall cost of the mission.

Moon over the Earth, as seen from the International Space Station, NASA photo

Energy Requirements

What makes the Skyhook concept work is energy exchange. Prior to a spacecraft docking at the lower end of the cable, the Skyhook uses its high efficiency low thrust onboard propulsion system to increase its orbital energy by raising it to a slightly higher orbit. When the arriving spacecraft docks with the lower end of the Skyhook, the Skyhook gives the extra energy of that higher altitude orbit to the arriving suborbital spacecraft, a spacecraft that does not have enough energy to remain in orbit on its own. The end result is that the arriving spacecraft gets an energy boost from the Skyhook that allows it to remain in orbit while the Skyhook gives up energy and drops to a slightly lower altitude, lower-energy orbit.

The same thing happens when a satellite or spacecraft is launched from the upper end of the Skyhook. As before, the Skyhook starts from a higher altitude, higher-energy orbit. When the satellite or spacecraft is released from the upper end of the cable, the Skyhook gives energy to that vehicle which places it in a higher-energy orbit. This addition of energy to the departing vehicle also means that the Skyhook now has less energy and as a result it is now in a lower altitude, lower-energy orbit.

Before another arriving spacecraft can dock at the lower end of the Skyhook, or before another satellite or spacecraft can be launched from the upper end, the Skyhook will need to use its onboard propulsion system to raise itself back to the original higher altitude, higher-energy orbit.

The energy for increasing its orbital altitude to a higher-energy orbit comes from either a solar powered ion propulsion system or a solar powered electrodynamic propulsion system on the Skyhook. The advantage of this over existing chemical propulsion systems is the greatly improved operating efficiency and reduced propellant needs of either of these propulsion systems. While these high efficiency low thrust propulsion systems cannot be used for a planetary surface to orbit launch system due to their high weight and low thrust, they are perfect for use on a non-rotating Skyhook due to their ability to gradually store up energy by raising the orbital altitude of the Skyhook between arriving and departing flights.

Launch Vehicles

Any orbital launch vehicle can fly to the lower end of a non-rotating Skyhook. Any orbital launch vehicle that flies to the Skyhook will increase its payload capacity and reduce its cost to orbit compared to flying directly to orbit on its own. This applies to all existing expendable launch systems. It also applies to air-launch to orbit launch systems such as Stratolaunch, air-breathers such as Skylon, and to reusable first stage launch systems such as the Falcon 9 and the one being developed by Blue Origin.

Chapter 15: The Basic Non-Rotating Skyhook

But this is the new ocean, and I believe the United States must sail on it and be in a position second to none.

John F. Kennedy, 1962

A non-rotating Skyhook is a vertically oriented, gravity gradient stabilized cable in orbit around the Earth, that serves as a bridge to space for suborbital spacecraft flying up from the Earth, and as a jumping off point for travel to the rest of the solar system. It is a bridge that can be built with existing technology that will make spaceflight so affordable that anyone who has the dream and the desire to experience it and explore it firsthand can go there.

The non-rotating Skyhook can also be thought of as an ion powered reusable second stage for suborbital launch vehicles coming up from the Earth, and as a reusable upper stage for spacecraft headed to a higher orbit, that never leaves Earth orbit.

The non-rotating Skyhook can also be viewed as the rope that hangs down from a treehouse that is used to lift things up and down.

The non-rotating Skyhook is also about making money. It is about safe, regular travel to an orbiting hotel for the price of a first class trip to Europe. It is about affordable to build satellite solar power stations that will beam clean, safe, affordable power down to Earth for a profit. It is about making asteroid mining so affordable that mining companies will be eagerly waiting in line to use it.

In short, the non-rotating Skyhook is a low-cost transportation system for hauling freight to Earth orbit and beyond, for a profit, and for a price that allows its customers to make money.

Some see private enterprise as a predatory target to be shot, others as a cow to be milked, but few are those who see it as a sturdy horse pulling the wagon.

Winston Churchill

The non-rotating Skyhook comes in many sizes. It can start as small as a 200-kilometer long footpath into space that can be built for less than the cost of a single Space Shuttle flight. This is a basic non-rotating Skyhook. It can also grow into a 4,000-kilometer long eight lane freeway that gets more affordable with every incremental increase in length, strength, and flight rate; improvements that are paid for with the profits it earns hauling the freight to Earth orbit and beyond during each phase of its development. This is a mature Skyhook.

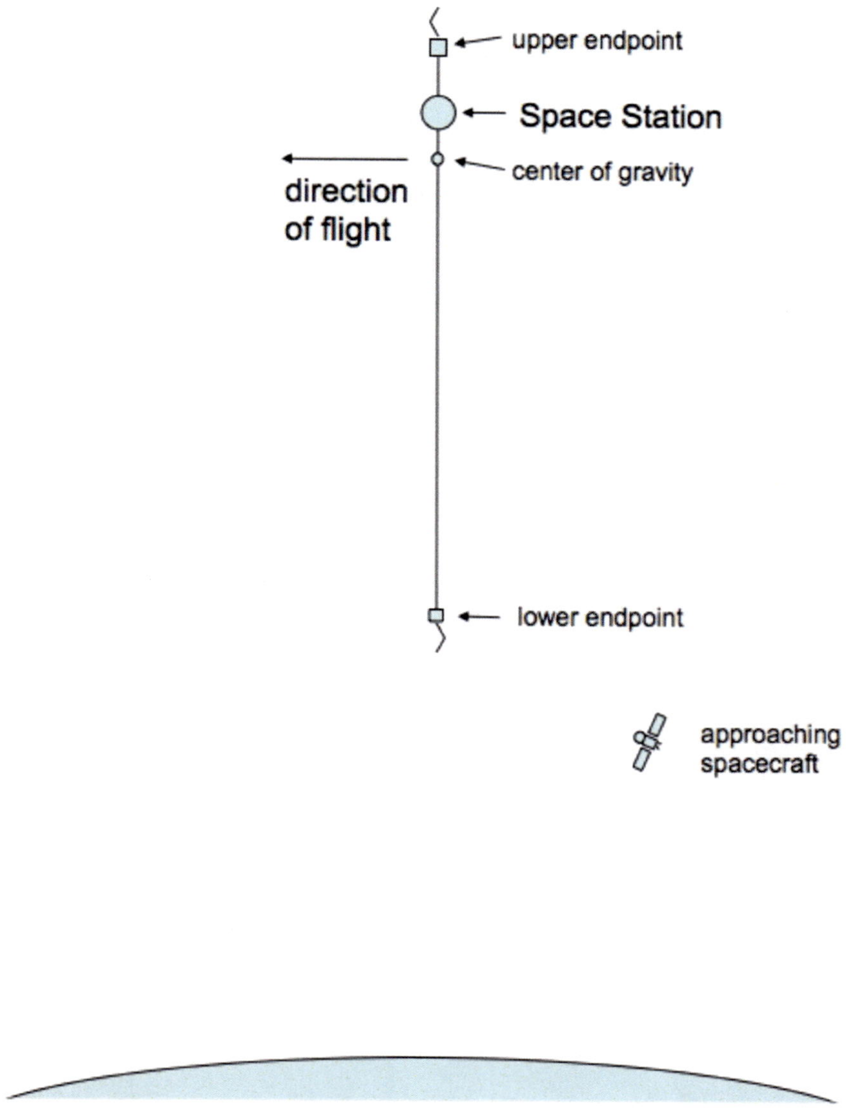

Basic Skyhook

Many scientists and engineers have investigated, validated, and added to the non-rotating Skyhook concept, including the fact that it can be built with existing materials and technology, and its cost-effectiveness.

In addition, NASA representatives who have reviewed this concept have described it as, "The first idea we have seen that offers a believable path to $100 per pound to orbit."

In its simplest form, a basic non-rotating Skyhook consists of a vertically oriented constant cross section cable with a small endpoint docking station attached at each end, and a central space station that is attached to the cable with a roller mechanism that allows it to move up and down the cable from one end to the other. In addition to the roller mechanism, the central space station has a habitation module (either rigid like the ones on the International Space Station, or inflatable as proposed by Bigelow Aerospace), two docking modules, two airlocks, a supply module, two remote manipulator arms, an attitude control system, a cooling system, an Earth return capable spacecraft that can serve as a lifeboat in case of an emergency, and a solar electric powered ion propulsion system. The purpose of this basic non-rotating Skyhook is to start getting real world operational experience with this concept with a minimum sized initial investment, while reducing the cost of getting into orbit at the same time. It is also designed so that it can grow into a longer and more cost-effective system.

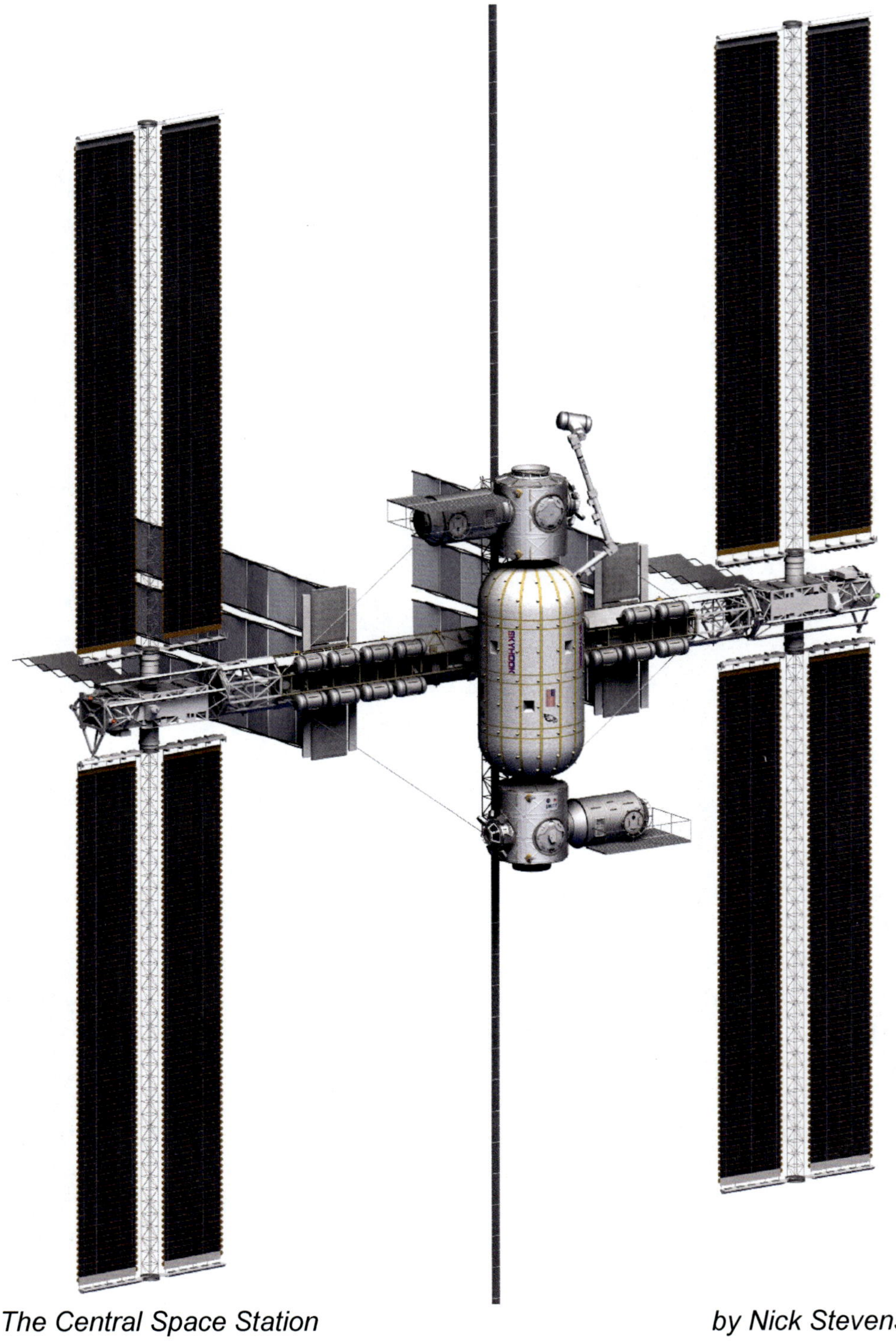

The Central Space Station *by Nick Stevens*

Four different overall lengths have been examined for this system: 200 kilometers, 260 kilometers, 320 kilometers, and 380 kilometers. Overall length refers to the overall length of the Skyhook cable, from the bottom of the lower end of the cable to the top of the upper end. Another term that will be used here is "useful length." Useful length refers to the distance between the center of gravity of the Skyhook and the end of the cable that a spacecraft is arriving at or departing from. Useful length is always shorter than the overall length. It is the useful length that determines the amount of change in velocity that the Skyhook makes possible. The useful length for the four different overall lengths and their corresponding velocity reductions are:

Overall length, km	Useful length, km	Velocity Reduction, m/s
200	150	260
260	200	343
320	250	425
380	300	505

How the Basic Non-Rotating Skyhook Works

Assume an initial 200-kilometer long vertically oriented cable with a mass of 5,600 kilograms that has a 2,000-kilogram endpoint station at each end, a 100,000-kilogram space station, and a 40,000-kilogram ion propulsion/power system designed for a flight rate of 12 arriving spacecraft per year. The baseline orbit for this basic Skyhook is an elliptical orbit that has a low point altitude of 350.6 kilometers, and a high point altitude of 445.3 kilometers. As with all Skyhooks, it is the center of gravity of the system that is in orbit. Since the speed of orbit decreases with increasing altitude, those parts of the Skyhook that are above the center of gravity are moving faster than orbital velocity for their altitude and want to move further away from the Earth, while those parts that are below the center of gravity are moving at less than orbital velocity for their altitude and want to move closer to the Earth. The end result is a vertically oriented cable that is under constant tension. Due to this tension, the cable/space station combination acts like a long thin rod with a weight that can be positioned anywhere along its length and the center of gravity of the system is determined by the vertical location of the space station on the cable.

Note The mass of the cable used in this example was calculated using the fiber properties for the T1000GB carbon fiber as shown on the TORAY Industries website (see reference 40). It utilizes a high temperature thermoplastic matrix for binding the fibers together and assumes a fiber volume of 65%. The cable is designed with a safety factor of 5 to 1, and assumes an arriving spacecraft mass of 14,800 kilograms, or 14.8 metric tons.

Preparing for an Arriving Spacecraft

In order to prepare the Skyhook for docking with a spacecraft coming up from the Earth, the Central Space Station will need to be positioned 170.3 kilometers up from the bottom end of the cable. This will put the center of gravity of the Skyhook at the 165.6-kilometer mark. This means that when the Skyhook is at the low point of its baseline orbit, the bottom end of the cable will be 185 kilometers above the surface of the Earth moving at 96.67% of circular orbit velocity for that altitude. That represents a velocity reduction to the launch vehicle of 260 meters per second.

The Rendezvous

For a spacecraft to rendezvous with the lower end of the Skyhook, its launch point will need to be in the orbital plane of the Skyhook at the time of take-off. This is normal for all launch vehicles that are flying to a specific orbit whether they are ground-launched or air-launched. In the case of ground-launching, either with or without a ground accelerator, it is entirely a matter of timing, of waiting for the position of the launch site to move under the flight path of the Skyhook as the Earth turns on its axis. In the case of an air-launch, it means flying the carrier aircraft to a location that will be underneath the flight path of the Skyhook at the time of launch.

A second requirement for this rendezvous to occur will be to time the moment of launch so that the spacecraft will be at the top of its flight path at the same time the lower end of the Skyhook is there. A real world analogy for this maneuver is a mid-air refueling operation between two aircraft. Both aircraft need to be at the same place, at the same time, moving in the same direction, and with the same speed for it to work. It is the same with the Skyhook.

A third requirement, that will need to be met for this rendezvous to occur, has to do with the elliptical baseline orbit of the Skyhook. The rendezvous will need to occur when the Skyhook is at the low point of its orbit. This requirement does not add to the technical difficulty of the rendezvous but it does limit the number times and locations that a rendezvous can occur in a day. For example: if the Skyhook takes two hours to orbit the Earth there will be 12 opportunities for a rendezvous in a day. The location of these rendezvous sites will be constantly changing because of the Earth rotating underneath the Skyhook and because the orbit of the Skyhook will constantly be changing as it stores up energy for the next rendezvous via its onboard propulsion system.

Hooking Up

When an arriving spacecraft docks with the bottom of the Skyhook it lowers the center of gravity of the system. If the arriving spacecraft has a mass of 14,800 kilograms, it will lower the center of gravity of this particular Skyhook by

15.6 kilometers: from 165.6 kilometers above the bottom end of the cable to 150 kilometers. This does not change the physical altitude of the Skyhook when this happens, as the lower end of the Skyhook is still 185 kilometers above the surface of the Earth. However, it does move the center of gravity of the Skyhook closer to the Earth. This changes the orbit of the Skyhook from the initial 350.6 x 445.3 kilometers high elliptical orbit to a 335 kilometers high circular orbit.

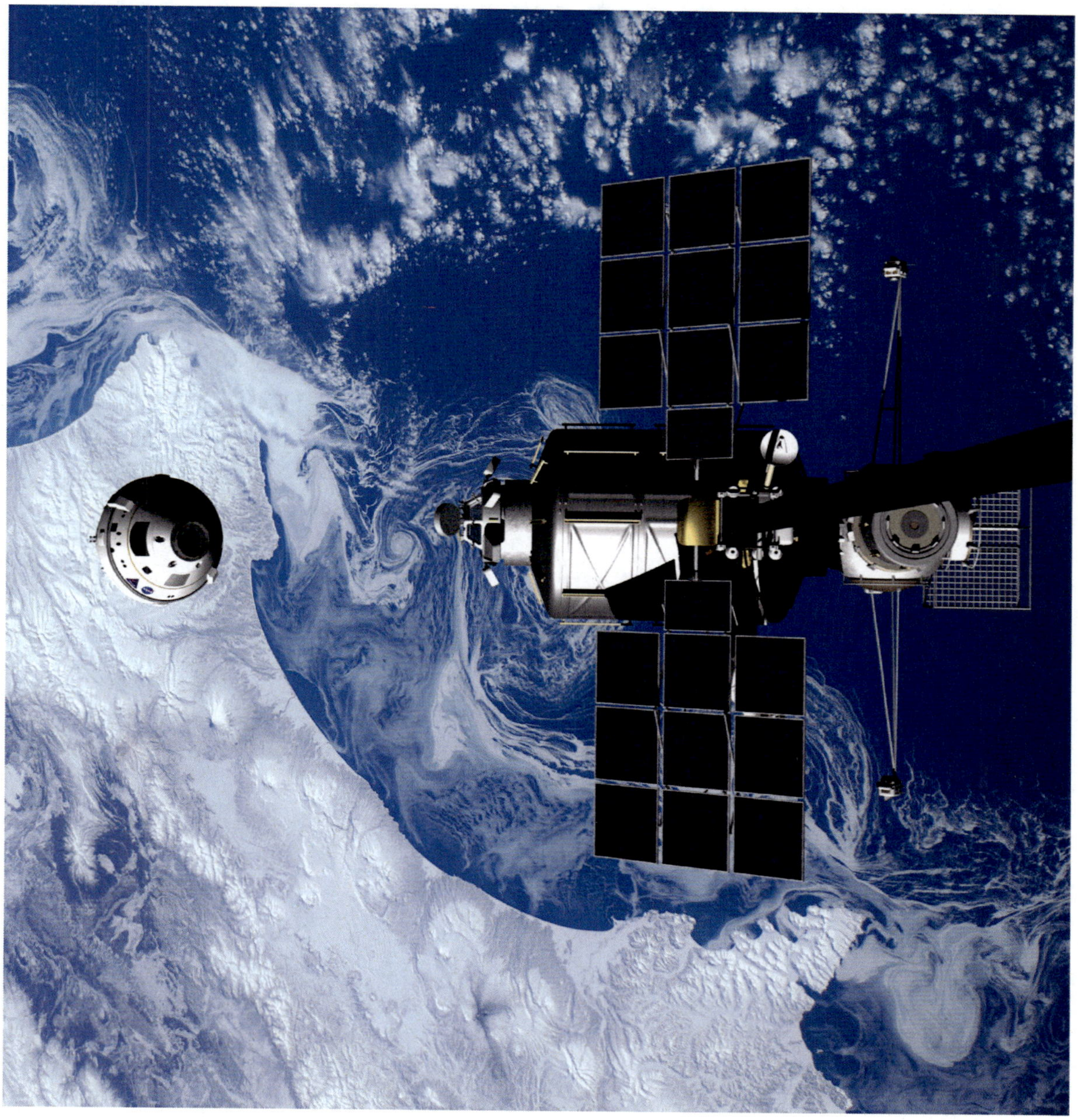

Orion approaching the Lower Endpoint Station *By Nick Stevens*

Orion docked with the Lower Endpoint Station, aka Skyhook Base 1 (SB1)
By Nick Stevens

Bringing the Spacecraft and Space Station Together

Once the arriving spacecraft is firmly attached to the Lower Endpoint Station, the Central Space Station begins to move down the cable. This lifts the Lower Endpoint Station up while at the same time lowering the Central Space Station down closer to the Earth. Since the Central Space Station is much more massive than the cable and endpoint stations, it is the cable with endpoint stations that changes altitude the most, while the Central Space Station changes altitude only a little. What is happening is that the center of gravity of the cable with endpoint stations is moving up while the center of gravity of the Central Space Station is moving down, but the altitude of the combined center of gravity of the total Skyhook system does not change. This means that the Skyhook remains in the same 335 kilometers high circular orbit while the Central Space Station and the Lower Endpoint Station with arriving spacecraft gradually come together and dock.

This picture, the one on the previous page, and the two on the following page that show the Central Station moving down the cable, are by Nick Stevens.

The Lower Endpoint Station with Orion spacecraft at the Central Space Station
By Nick Stevens

The Orbit Raising Maneuver

While the Central Space Station is gradually moving down the cable towards the Lower Endpoint Station, the ion propulsion system is activated in order to start raising the orbital altitude of the Skyhook. Once the Central Space Station has arrived at the Lower Endpoint Station and the crew change has occurred, the Lower Endpoint Station with attached spacecraft is lowered back down approximately 45 kilometers. Since the departing spacecraft is moving at less than orbital velocity while hanging on the cable at this altitude, no retro-rocket burn is required for the return to Earth, and the release of the spacecraft will need to occur when the Skyhook is in the proper position over the Earth for the departing spacecraft to arrive at its planned landing site. After the departing spacecraft has left, the ion propulsion system will continue thrusting until the Skyhook is back in its baseline elliptical orbit for the next arriving spacecraft. The ion propulsion system will use 87 kilograms of propellant for this maneuver. *For a short video that shows how the Skyhook works see reference 63.*

Growing the Skyhook

As demand for flights to the Skyhook increases, the Skyhook will be progressively lengthened so as to further reduce the velocity that a launch vehicle needs to achieve in order to get to the lower end. This will increase the useful payload of the spacecraft and further reduce the cost of getting to orbit. A second ongoing upgrade will involve increasing the strength of the cable so that larger spacecraft with larger payloads will be able to use the Skyhook. A third ongoing upgrade will consist of gradually enlarging the ion propulsion system so that the flight rate can be increased. These three ongoing upgrades will be accomplished by filling any unused cargo capacity in an arriving spacecraft with materials for growing the Skyhook, and by flying dedicated cargo missions. All of this will grow the Skyhook into a longer, stronger, higher flight rate system that will continue to reduce the cost of getting into orbit with each addition.

Best Orbit for a Skyhook

The best orbit for a Skyhook depends on both the type of launch system that is being used to fly to the bottom of the Skyhook, and the flight rate. If the launch vehicle is being air-launched, then any orbit is possible as the carrier aircraft will be able to fly to wherever it needs to be in order to facilitate a rendezvous between the launch vehicle and the bottom end of the Skyhook.

If the launch vehicle is taking off from a fixed ground site, either with or without a ground accelerator, and the flight rate is low, then waiting for the Skyhook to move into position for a launch is an acceptable answer. However, as demand for flights increase due to the reduced cost of going to orbit, the limited number of launch opportunities from a fixed site will become an issue.

Therefore, when launching from a fixed launch site with a higher flight rate, the best orbit for the Skyhook will be an equatorial orbit, and the best location for the launch site will be somewhere on or near the equator.

Navigating a Skyhook

Navigating a Skyhook and controlling the location of the low point of its orbit, its orbital eccentricity, and its average orbital altitude will be very different from any other Earth orbiting satellite or spacecraft that has been put into orbit before. The primary difference between a Skyhook and previous orbital vehicles is that the Skyhook will be under almost continuous acceleration and as a result the location of the low point of its orbit, its orbital eccentricity, and its average altitude will be in a near constant state of change.

For a Skyhook in an orbit that is inclined to the plane of the equator, there will also be the additional issue of nodal precession and how to manage the change in the orbital plane this causes so that the Skyhook will be at the desired location at the right time for whatever flight activity is planned.

Another significant issue regarding the navigation of a Skyhook is the low thrust to weight ratio of the Skyhook's propulsion system and the resulting low rate of acceleration. It is due to this low rate of acceleration that every spacecraft arrival at the lower end of the Skyhook, every return to Earth from the lower end of the Skyhook, and every launch to a higher orbit from the upper end will need to be planned many months in advance and coordinated with other planned arrivals, departures, and upper endpoint launches in order to make sure the Skyhook is in the right place at the right time so as to fulfill each specific mission and to maximize its utilization.

Control of all these factors is made possible by the Skyhook's propulsion system and by planning in advance so as to minimize the total change in velocity required by the Skyhook to accomplish the various missions. Some of the possible control methods for accomplishing this are:

- vectoring and varying the thrust of the ion propulsion system;
- pausing for a period of time in a specific orbit that has an orbital period or nodal precession rate that allows the Skyhook to end up in the desired location at the desired time; and,
- timing the release of refuse from the lower end of the Skyhook so that the resulting change in orbit assists in flying the Skyhook to the desired location at the desired time.

Note Taking out the trash is something that needs to be done in every structure that humans inhabit. The same holds true for a space station or a Skyhook. On the International Space Station, the trash is placed in an empty unmanned cargo spacecraft, which is then deorbited and allowed to burn up in the Earth's atmosphere. On a Skyhook, the trash will be attached to the Lower Endpoint Station, lowered down to the maximum extension of the cable so as to maximize the momentum recovery, and released to burn up in the atmosphere.

Imagine the View!

This picture shows the lower docking module of the Central Space Station as it approaches the Lower Endpoint Station. When the Central Space Station gets close to the Lower Endpoint Station it is far enough away from the Skyhook's center of gravity that the station experiences a small amount of gravity. This is why the astronaut is standing and why there is a porch attached to the airlock module. At the moment of this photo, the Skyhook is in the Earth's shadow and the Station's external floodlights are turned on. The gold colored panels behind the astronaut are the backsides of the solar arrays lit up by the floodlights. *By Nick Stevens*

Jupiter, NASA photo

Skylab, NASA photo

Chapter 16: A Mature Non-Rotating Skyhook

As time progresses and the basic non-rotating Skyhook has its cable lengthened and strengthened, and its flight rate increased so as to serve a growing market, it will eventually become like the mature Skyhook shown here.

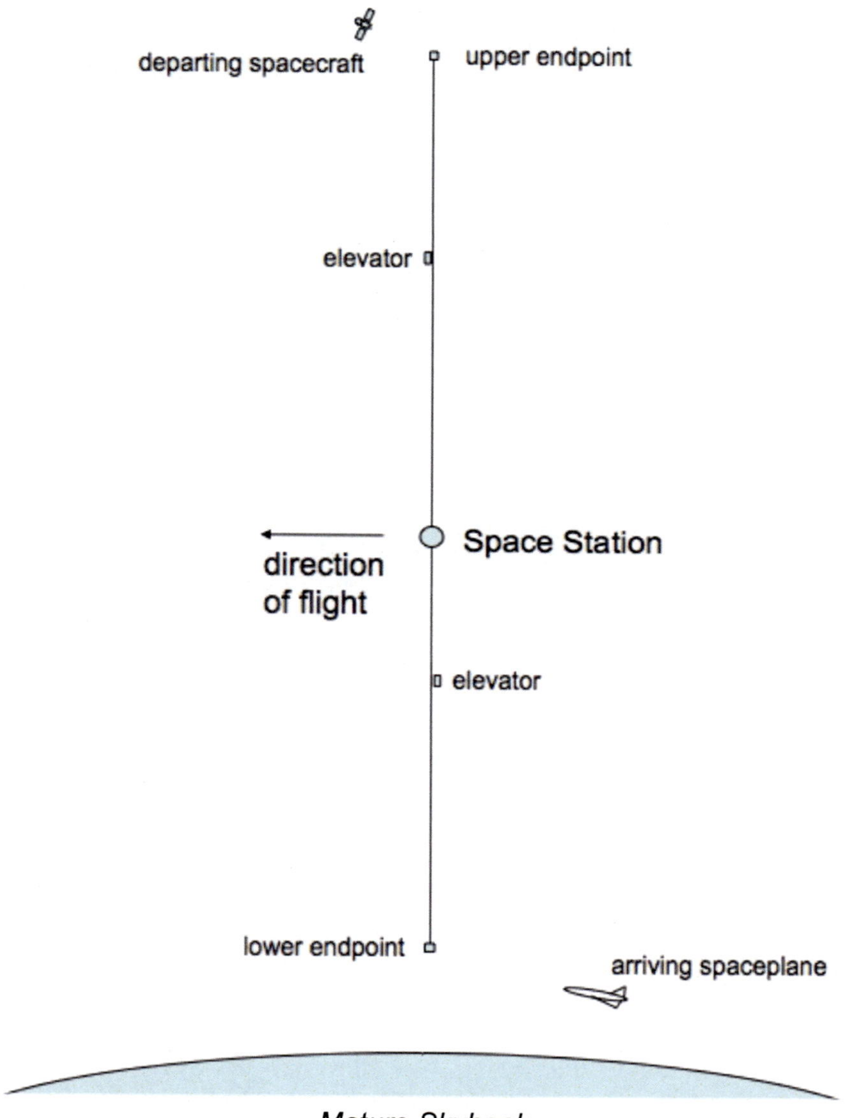

Mature Skyhook

This Skyhook has both a dedicated upward and downward pointing cable, as well as a pair of elevators for travel between the Central Space Station and the two endpoint stations. The upper cable is designed to balance the weight of the lower cable and to be long enough that the Upper Endpoint Station will be moving at just short of escape velocity for its altitude.

At this point the mature Skyhook will be a terminal for suborbital launch vehicles flying between the Earth and the Lower Endpoint Station, as well as a terminal for spacecraft going to, or arriving from, higher orbit, lunar orbit, or the Outpost Space Station at the L1-L2 Earth-Moon Lagrange Point, at the Upper Endpoint Station. The Central Space Station in the middle will also serve as a terminal for space taxies traveling back and forth between the Skyhook and other space stations and orbiting hotels in nearby orbits.

Orion spacecraft launching from the Upper Endpoint Station heading for higher orbit. *By Nick Stevens*

At what overall length it will be necessary or best to switch over from the basic Skyhook configuration to the mature Skyhook configuration with its segmented and step-tapered upper and lower cables and elevators is currently unknown. Most likely the answer will not be known until a few years of experience have been acquired operating a basic Skyhook.

The amount of velocity reduction for an Earth arriving launch vehicle made possible by a mature Skyhook will depend on the useful length of the lower cable. The longer it is the greater the velocity reduction. Unfortunately, longer also means heavier and more expensive, which is why starting small and growing with the market is so important.

Useful length of the lower cable, km	Velocity reduction, m/s
400	663
500	814
600	960
700	1,100
800	1,237
900	1,368
1,000	1,495
1,200	1,735
1,400	1,965
1,600	2,178
1,800	2,378

The Upper End of a Mature Skyhook

The upper end of a mature Skyhook is designed to be moving at slightly less than escape velocity for its altitude. What this means is that a spacecraft departing from the upper end of the Skyhook can, with the use of only a very small amount of onboard propellant, be headed for geosynchronous orbit, on a free return trajectory to the Moon, on its way to one of the Lagrange Points, or headed for a near-Earth asteroid.

To put all this into perspective, consider the Saturn V rocket that was used to fly to the Moon in the late 1960s and early 1970s. The Saturn V had a take-off mass of approximately 2,974 metric tons. It used 2,723 metric tons of propellant and 189 metric tons of launch vehicle to place 62 metric tons of payload onto a Lunar Transfer Orbit (LTO). That works out to a total of 43.7 metric tons of propellant and 3.0 metric tons of launch vehicle per metric ton of payload delivered to Earth escape velocity.

Now compare the Saturn V to a mature non-rotating Skyhook launch system.

A highly reusable winged suborbital spaceplane powered by a combination air-breathing/rocket motor propulsion system flies to the bottom end of a mature non-rotating Skyhook. The spaceplane is carrying a cargo that is bound for the Moon. The spaceplane could be air-launched, ground accelerator launched, or it could have afterburning turbine engines and take-off on its own. Whichever it is, the spaceplane is a single stage vehicle that is completely reusable and operates like a modern-day airliner.

One design study of such a launch system that was performed in the mid-1990s used a horizontal take-off winged spaceplane with an afterburning turbine/rocket/ramjet/scramjet propulsion system for flying to a mature Skyhook that had a lower useful cable length of 1,800 kilometers. That spaceplane had a take-off mass of 165 metric tons, used 94 metric tons of propellant, and delivered 16 metric tons of payload to the lower end of the Skyhook. That comes out to:

Propellant fraction	56.9%
Empty weight fraction	33.3%
Gross payload fraction	9.8%

Upon arrival at the lower end of the Skyhook, the cargo from the spaceplane is transferred to the Skyhook and the spaceplane returns to Earth with a load of returning passengers and cargo. Next, the lunar-bound cargo is placed on the elevator that will carry it up the tether to the Central Space Station located near the middle of the Skyhook. From there it is transferred to the elevator that goes up to the Upper Endpoint Station.

The ion propulsion system on the Skyhook uses approximately 180 kilograms of propellant per metric ton of passengers and cargo when moving them from the lower end of the Skyhook to the upper end. That means that of the original 16 metric tons that arrived at the lower end of the Skyhook, approximately 13.5 metric tons will be delivered to the upper end of the Skyhook assuming that the propellant for moving this cargo up the cable was included with this delivery. It is also important to realize that by moving the cargo to the Upper Endpoint Station it has effectively been accelerated to the speed of a Lunar Transfer Orbit.

The end result is that the total amount of propellant used per ton of payload delivered from the surface of the Earth to the Upper Endpoint Station with this spaceplane/Skyhook system is approximately 7.1 metric tons. The amount of hardware per ton of payload that is used up by this system is zero.

In Summary

Saturn V to LTO	
Tons of propellant per ton of payload	43.7
Tons of hardware per ton of payload	3.0
Total	46.7
Spaceplane to Skyhook, Skyhook to LTO	
Tons of propellant per ton of payload	7.1
Tons of hardware per ton of payload	0.0
Total	7.1

Note The 7.1 tons of propellant per ton of payload to LTO for this spaceplane/Skyhook launch system will become even less if the spaceplane is either air-launched or ground accelerator launched. Either of these additions to the launch system will reduce the speed required of the spaceplane thereby reducing its propellant fraction. Either of these additions will also eliminate the weight of the afterburning turbine engines, which will reduce the empty weight fraction of the spaceplane. These weight reductions will increase the payload fraction of the spaceplane, which will further reduce the cost.

Continuing the Journey

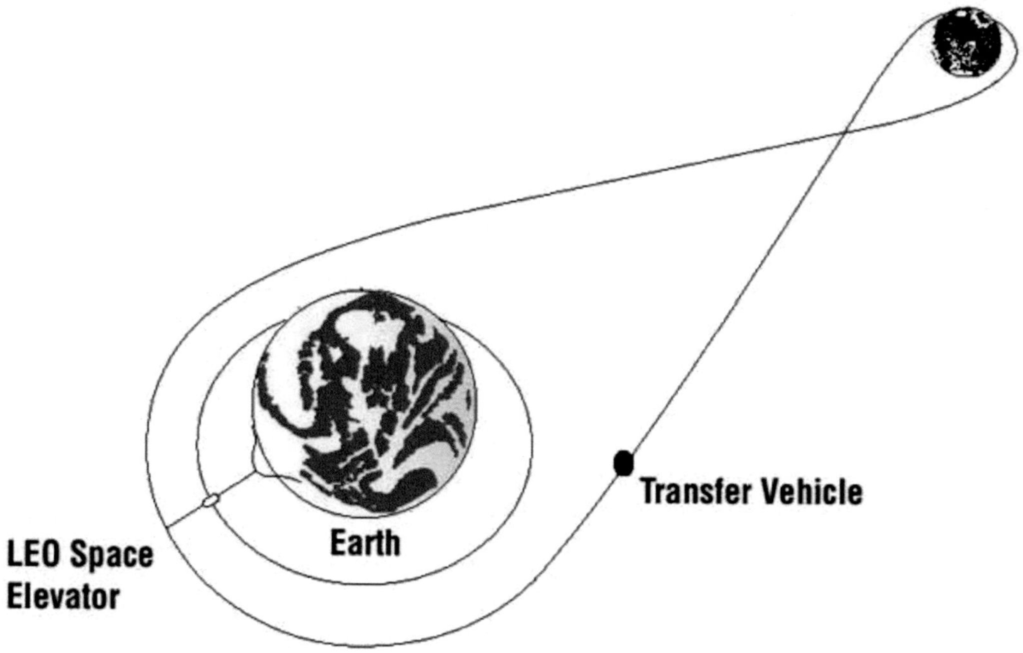

An Earth-Moon transfer via a Non-rotating Skyhook (referred to as a LEO Space Elevator in this NASA drawing)

At the Upper Endpoint Station, the lunar-bound cargo will be loaded onto a small reusable Earth-Moon spacecraft that will take it either to lunar orbit or to the outpost space station at L1 or L2. This Earth-Moon spacecraft will use a small amount of onboard rocket propellant when it departs the upper end of the Skyhook, and will then fire its rocket motor a second time to slow down to lunar orbit velocity when it arrives at the Moon. All of the propellant used by this spacecraft will come either from the Moon or from an asteroid that has been moved into orbit around the Moon.

Upon arrival in lunar orbit, the Earth-Moon spacecraft will rendezvous and dock with a Lunar Orbit Space Station and Propellant Depot. Here it will refuel and take on a load of returning passengers and cargo before heading back to the upper end of the non-rotating Skyhook. The lunar-bound cargo will then be transferred to a reusable single stage Lunar Lander that will take it to the lunar surface.

The Altair Lunar Lander as envisioned in 2010, NASA

Notice how all the flight hardware used in the Skyhook transportation system is reusable as compared to the Saturn V where everything was expendable. Also notice the tremendous reduction in the amount of propellant that is required and how this reduces the overall size and cost of all the flight vehicles. All of these changes, the 100% reusability of the system, the refueling station in lunar orbit, the much smaller size, the much lower propellant fraction and the much higher payload fraction of all the flight vehicles, these are the changes that will allow spaceflight to become mass-market affordable.

The Earth-Moon Spacecraft

NASA is currently developing a reusable Earth-Moon/Earth-Asteroid spacecraft called the Orion spacecraft. Combine this new spacecraft with a non-rotating Skyhook and an affordable transportation system for traveling to lunar orbit, the Earth/Moon Lagrange points, and the near-Earth asteroids will exist. Add a single stage reusable Lunar Lander and an affordable transportation system for going to the lunar surface will exist. Once access to Lunar and/or near-Earth asteroid materials for propellant and shielding exists, the building of an affordable, well shielded, rotating Mars cycler spacecraft as proposed by Buzz Aldrin becomes possible. Once this is built, regular safe and affordable interplanetary spaceflight to Mars, the moons of Mars, as well as Ceres and all the other asteroids between Mars and Jupiter, becomes possible.

Orion spacecraft, NASA

Note The reason for limiting the speed of the Upper Endpoint Station to less than escape velocity is to simplify the rescue and recovery of any astronaut who accidently falls off the Upper Endpoint Station while working outside in a spacesuit. Due to the fact that both the Upper Endpoint Station and the Lower Endpoint Station are moving either too fast or too slow for orbit at their respective altitudes, people at those stations will experience gravity. Not full Earth gravity, but some degree of gravity, the exact amount being dependent on the length of the cables. As a result, it will be possible for an astronaut working outside at either of these stations to fall off. By limiting the speed of the Upper Endpoint Station to less than escape velocity, the rescue of an astronaut in this situation becomes a relatively simple operation.

Chapter 17: Combination Launch Systems

Man will not always stay on Earth; the pursuit of light and space will lead him to penetrate the bounds of the atmosphere, timidly at first, but in the end to conquer the whole of solar space.

Konstantin Tsiolkovsky

There are many possible combination launch systems based on these five concepts. As to which is the absolute best - that is currently unknown. Most likely there are multiple best answers and it will take time, operational experience, and free market competition to sort it all out. It is also likely that the best answer will change as the market grows and develops.

Air-Launched Reusable First Stage/Expendable Upper Stage Launch Vehicle with Basic Skyhook

One possible combination launch system would be a large subsonic carrier aircraft for air-launching a horizontal landing rocket powered reusable first stage with expendable upper stage launch vehicle. The carrier aircraft to position itself for launch so that the reusable first stage can glide back to the airport the carrier aircraft took off from. Once this launch system is operational and has proven itself launching satellites and space station resupply flights, a possible upgrade to the system would be to add a basic Skyhook to the system and start flying passengers and cargo to the Skyhook.

The reduction in the speed required for orbit made possible by air-launching the launch vehicle is approximately 1,100 meters per second. Once the basic non-rotating Skyhook is added to the system there will be an additional velocity reduction of between 260 meters per second and 500 meters per second, depending on the initial length of the Skyhook. Once the Skyhook has grown to a fully mature system the total speed reduction from air-launch and the Skyhook will be over 3,400 meters per second.

It is also important to remember that the addition of wings to the launch vehicle that will be required for air-launch will also increase the speed required for orbit by approximately 400 meters per second. Therefore, the speed required for flying to the lower end of the Skyhook by the rocket in the fully mature version of this launch system is approximately 6,100 meters per second.

VALT Launched Vertical Landing Reusable First Stage/Expendable Upper Stage Launch Vehicle with Basic Skyhook

Another possible combination system would be to use a 600 MPH Vertical Accelerator Launch Tower for launching a rocket powered vertical landing reusable first stage with expendable upper stage launch vehicle. The Vertical

Accelerator Launch Tower to be located towards the western edge of whatever landmass it is located on so that the vertical landing reusable first stage will be able to land on a ground-based landing pad instead of a barge out at sea. These landing pads to be little more than a collection of concrete pads arranged in an arc downrange of the launch tower so as to provide a landing site for the reusable first stage regardless of what orbit it is flying to. The return of the reusable first stage to the launch tower will be performed using either a tractor-trailer or by refueling the reusable first stage and flying it home.

Like the previous example, a possible initial market for this launch system is launching satellites into Earth orbit, and for crew and cargo flights to a space station. Another possible market would be the launching of suborbital tourist flights using the ground accelerator, the reusable first stage, and a reusable suborbital passenger spacecraft.

Once this part of the transportation system is operational, the next step would be the addition of a basic non-rotating Skyhook to the system and to begin transporting passengers and cargo to the lower end of the Skyhook.

One possible disadvantage of this system will be the amount of time it will take for the Skyhook and the ground accelerator to come back into alignment with each other for another launch. In the beginning, when flight rates are low this will not be a problem, but as the flight rate increases, it will place a significant limit on the system. A solution to this problem would be to place the ground accelerator on the equator and to place the Skyhook in an equatorial orbit.

The velocity reduction made possible by this system is 800 to 900 meters per second for the ground accelerator, and 260 to 500 meters per second for the initial Skyhook, depending on its length. Once the Skyhook has grown to a fully mature system the total speed reduction from VALT and the Skyhook will be over 3,100 meters per second. Therefore, the speed required for flying to the lower end of the Skyhook by the rocket in the fully mature version of this launch system is approximately 6,000 meters per second.

Trackless Ground Accelerator with a Winged Ramjet/Rocket Reusable First Stage/Expendable Upper Stage Launch Vehicle with Skyhook

A third possibility would be to build an all-orbit capable trackless ground accelerator for launching a winged ramjet/rocket powered reusable first stage with expendable upper stage launch vehicle. Due to its low-cost to build, this ground accelerator could be located either on the equator or at a non-equatorial site depending on whatever works best for the initial customer base. As with the other two options, the initial market for this launch system will most likely be for launching satellites and for crew and cargo flights to a space station.

The next step would be the addition of a basic non-rotating Skyhook; most likely in an equatorial orbit so as to maximize launch opportunities. If the initial ground accelerator is built at a non-equatorial site, a second trackless ground accelerator would need to be built on the equator for flying to the Skyhook.

Basic Skyhook with Existing Launch Vehicles

A fourth possible way of getting started would be to place a basic non-rotating Skyhook in an equatorial orbit and to launch a reusable first stage/expendable upper stage launch vehicle from an equatorial launch site such as Guiana Space Center in South America to the Skyhook. The reusable first stage of the launch vehicle to land either at a ground-based landing pad near the launch site or on a barge out at sea.

These are just a few of the combination launch systems that are possible using these five concepts: ground accelerators; air-launch; combination air-breathing and rocket motor propulsion systems; reusable first stage launch vehicles; and, non-rotating Skyhooks. There are other possible combinations. What makes them all work and more affordable than existing launch systems, is the reduced velocity requirement for the launch vehicle, and the fact that they are all based on existing technology. The reduced velocity requirement allows the launch vehicle to carry a larger payload, and once the velocity reduction is large enough, to make the launch vehicle 100% reusable. That these concepts are all based on existing technology allows them to be built for a price that can be predicted and planned for in advance. This predictability allows for a realistic business plan to be put together. Now, instead of waiting for some new magic technology to come along that will make it all possible, the problem has become finding the combination launch system with the lowest start-up cost, the least risk, that works with the existing launch market, and that can grow and evolve as the market grows and evolves.

It is no longer necessary to wait for the magic to arrive anymore – it has arrived.

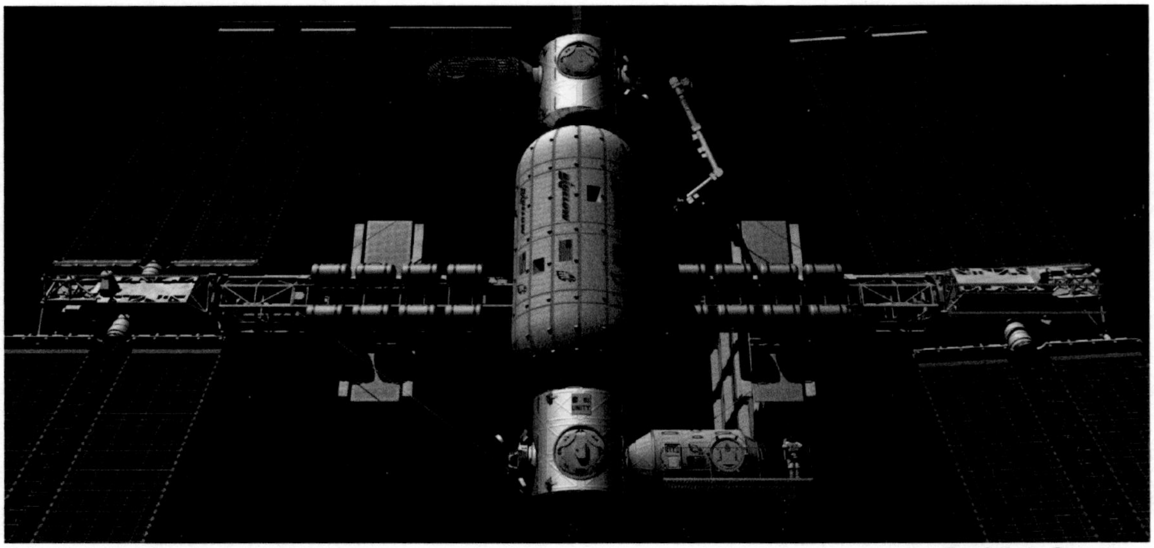

By Nick Stevens

Chapter 18: The Near Future

I searched along the changing edge
Where sky-pierced now the cloud had broken.
I saw no bird, no blade of wing,
No song was spoken.
I stood, my eyes turned upward still
And drank the air and breathed the light.
Then, like a hawk upon the wind,
I climbed the sky, I made the flight.

Inspired by Elizabeth J. Buchtenkirk

Taking off from Spaceport America in New Mexico, we began our flight to the lower end of the non-rotating Skyhook that orbits the Earth. Upon reaching 40,000 feet of altitude, on the proper heading, at the right location, and at the right time, our spaceplane is released from the subsonic carrier aircraft. After a few moments of free fall where everyone is held in their seat by their seat harness, our pilot throttles up the ramjets and the feeling of gravity and acceleration returns. A moment later the ducted rocket motors are ignited for a brief period in order to get us through the sound barrier and up to Mach 1.2, a speed where the ramjets really start to perform. A feeling of elation fills me as I look at the outside view shown on the video monitor in front of me, a feeling that grows and grows as the knowledge that we are going to ride this beauty all the way into space really starts to sink in!

The spaceplane for this flight is a winged, air-launched, horizontal landing, hypersonic aircraft, that has a dual-fuel/liquid oxygen powered Rocket Based Combined Cycle propulsion system. This propulsion system is a combination system that consists of ramjets, scramjets, and multiple ducted rocket motors, that all use a shared flow path.

Onward we go. We continue to accelerate and climb, and the sky starts to darken. Upon reaching Mach 5, the ramjets reconfigure themselves for scramjet flight and continue accelerating to Mach 8. It is a smooth transition, not even noticeable except for the announcement by the pilot. The acceleration continues but gradually gets less. At Mach 8 we are so high and the air is so thin that the scramjets no longer have enough thrust to continue accelerating, so we began powering up the ducted rocket motors for the second time. There is a bit of a rumbling vibration as the many small rocket motors that are distributed throughout the flow path of the ramjet/scramjet are ignited in a rolling start and then the acceleration increases as the pilot throttles them all up. As our speed increases, the pilot pulls back on the stick and the nose of our spaceplane pitches to a higher angle. Now we begin leaving behind the small amount of atmosphere that still exists at this altitude as we begin to really head for space.

Looking at the video monitor, I see that the sky is getting darker by the moment and the curve of the Earth is getting more and more pronounced.

NASA photo

The Mach meter keeps climbing. At approximately Mach 15, the rocket motors on the spaceplane are shut down and shortly thereafter the payload bay doors open and our little spacecraft is released to fly on its own. Now that we are clear of the spaceplane we can see out the windows of our spacecraft and the view is mesmerizing. At the same time, the video monitor in front of me switches from "outside view" to "navigation mode" so that it is possible to see our flight path and our position relative to the lower end of the Skyhook. Once our GPS system has confirmed our position and flight path, the pilot uses the spacecraft's onboard rocket motors for the final boost to get us up to rendezvous speed for our meeting with the lower end of the Skyhook. It's a relatively short burn with mild acceleration, but it does the job. It is also exciting to see how this changes our flight path on the video monitor and how close it gets us to where we will need to be for a successful rendezvous.

As we coast in zero-gee on our way up to the Skyhook there are additional small course corrections that adjust out flight path until it shows that we will arrive within inches of where we will need to be for the crane on the lower endpoint station to capture our spacecraft. The precision is amazing and we get to witness the entire process on the video monitor. Just as we arrive at the lower endpoint station I also got to see the crane arm move by my window during the capture process. A few minutes later the crane swings our spacecraft into position and our docking collar is attached to the Skyhook. Shortly thereafter the hatch is opened and we get to climb into the lower endpoint station.

There is a small amount of gravity at the lower end of the Skyhook, which helps to settle the stomachs of those who find zero-gee uncomfortable. It is not full Earth gravity, but it is enough. There is also a calming quality to it that gives reality to the experience. It also gives me a chance to reconnect with the world after the excitement of the launch. Standing here in the light gravity looking out the window at the Earth below and the universe above, it starts to really sink in; I am here, in the heavens, standing at the edge of the universe.

NASA photo

After separating from the spacecraft, the spaceplane re-enters the atmosphere, possibly using the same technique that Burt Rutan developed for SpaceShip One and SpaceShip Two. After slowing down to Mach 1.5, the spaceplane will then re-start its ramjets and fly to a landing site where it will either be picked up by the carrier aircraft for the return trip back to Spaceport America, or if at a spaceport, will be serviced and prepared for another boost to the Skyhook.

The spaceplane will most likely fly the boost phase of the launch unmanned and under the control of the pilot in the spacecraft in the payload bay, and on autopilot for its reentry and landing.

On October 3, 1967, the X-15A-2 reached a speed of 6,630 feet per second (Mach 6.70, 4,520 MPH). The X-15A-2 was a stretched X-15 with external drop tanks and a special ablative coating for dealing with the high temperatures that occur at these speeds. The X-15A-2 was designed for speeds up to Mach 8 and as a carrier aircraft for testing a supersonic combustion ramjet (a mockup of the scramjet can be seen in the photograph under the ventral fin at the rear of the aircraft). The pilot for this flight was Major William J. Knight. NASA, the U.S. Air Force, the U.S. NAVY, and North American Aviation jointly sponsored the X-15 program. U.S. Air Force photo

Chapter 19: Speed Reductions and the Rocket Equation

The Rocket Equation, also known as the mass ratio equation, is used to calculate the amount of propellant a rocket needs to carry to reach a certain speed. If speed is thought of as an analog to distance, it can also be thought of as how much gas a car needs to carry to go a certain distance. The more gas the car can carry, the further it can go. However, there is a limit on how much gas a car can carry, which puts a limit on how far it can go without refueling. The same applies to a rocket, the more propellant it has the faster it goes. However, just like a car, there is a limit on how much propellant the rocket can carry, which puts a limit on its speed.

The importance of this is that the speed required for reaching low Earth orbit is one of the numbers used in the rocket equation to calculate the amount of propellant the rocket needs to carry, which then determines how much weight is leftover for the empty weight of the rocket and the payload. It is also important to realize that the speed required for orbit varies depending on the type of launch vehicle being used. A vertical take-off, two-stage to orbit launch vehicle with no strap-on boosters like the Titan 2, has a speed required for low Earth orbit of approximately 9,100 meters per second. A launch vehicle like the Space Shuttle with side mounted boosters and a side mounted Shuttle has a speed required for orbit of approximately 10,000 meters per second. A launch vehicle that uses wings to help it fly into orbit like Pegasus has a total speed required for orbit of approximately 9,500 meters per second.

Another necessary number used in the rocket equation is the efficiency of the rocket motor. Think of this number as the number of miles per gallon that a car gets. With a rocket motor this term is called Specific Impulse and it measures the amount of time, in seconds that a rocket motor with a given amount of thrust, will run on a given amount of propellant. The commonly used abbreviation for specific impulse is Isp.

If these two numbers, speed required for orbit, and specific impulse are put into the rocket equation, it then becomes possible to calculate the amount of propellant the rocket needs to carry to reach that speed. If the velocity reduction value for air-launch, a ground accelerator, a Skyhook, or a combination of those is subtracted from the speed required, it then becomes possible to determine the reduction in propellant required, the increase in payload capacity, and how this changes the cost of going to orbit.

Example 1 The best way to show how this works is with an example that uses a real launch vehicle whose weight and performance numbers are known. For this example, the Titan 2 GLV launch vehicle was selected. The Titan 2 GLV was used to launch the Gemini spacecraft into low Earth orbit back in the 1960s. The Titan 2 was a basic two-stage expendable launch vehicle where the second stage was stacked on top of the first stage. It did not use any strap-on boosters or side mounted propellant tanks. There was also no payload fairing or escape tower to jettison, all of which noticeably simplifies the example.

Gemini spacecraft on top of a Titan 2 GLV launch vehicle, NASA photo

---

1st stage of the Titan 2

empty	6,736 kg
propellant	111,130 kg
stage total	117,866 kg
Isp	296 seconds

2nd stage of the Titan 2

empty	2,404 kg
propellant	26,535 kg
stage total	28,939 kg
Isp	316 seconds
Gross Payload to LEO	3,600 kg
Total	150,405 kg

Using these numbers and the rocket equation it is then possible to calculate the amount of speed that each stage provided and the total amount of speed that the Titan 2 needed to reach low Earth orbit.

Stage 1	3,895 m/s
Stage 2	5,234 m/s
Total	9,129 m/s

Note The speed for low Earth orbit is approximately 7,800 meters per second. The difference between this value and the total speed of the Titan 2 launch vehicle is the amount of speed that is lost overcoming the force of gravity and the aerodynamic drag of the launch vehicle as it climbs to orbit. The exact amount of speed loss is different for each launch vehicle and depends on how quickly the launch vehicle accelerates and how much drag it has.

There is one more item that will need to be added to this example before it is possible to examine the impact of combining a ground accelerator and a Skyhook with the Titan 2, and that is a cargo spacecraft that will sit on top of the Titan 2 launch vehicle. It is this cargo spacecraft that will carry the useful payload on the last leg of its journey to the space station. Unfortunately, no cargo spacecraft was ever developed for this launch vehicle so it will be necessary to guestimate one based on other existing cargo spacecraft. There are five: Progress (Russia); Dragon by SpaceX; HTV (Japan); Cygnus by Orbital; and, ATV (Europe). The useful payload mass of all of these cargo spacecraft is approximately 35% of their launch mass.

Cargo Spacecraft for the Titan 2 Launch Vehicle Based on all of this information, it will be assumed that the cargo spacecraft for the Titan 2 launch vehicle will have a launch mass of 3,600 kilograms and a useful payload mass to a space station of 1,260 kilograms. Due to the lack of requirement for deorbit propellant when departing from a Skyhook equipped space station, the baseline useful payload of this cargo spacecraft will be increased to 1,400 kilograms when flying to a Skyhook. Reductions to the speed required for orbit that come with adding a ground accelerator and a Skyhook to the system will be adjusted for by reducing the amount of propellant carried in the second stage. This reduction in propellant mass will then be added directly to the useful payload mass of the cargo spacecraft. For this example, it will be assumed that there is enough excess volume in the cargo spacecraft to accommodate this additional payload.

Progress spacecraft near the International Space Station, NASA photo

As already shown, the speed required for orbit by the Titan 2 launch vehicle is 9,129 meters per second. This is a real world number based on what the Titan 2 actually did. If either a 600 MPH mountain slope ground accelerator or a 600 MPH VALT style ground accelerator is used to launch the Titan 2 from a 12,000-foot elevation, then the speed required for orbit can be reduced by 800 meters per second. The reason for choosing a ground accelerator instead of an air-launch for this example is due to the fact that the Titan 2 did not have wings and this saves the trouble of trying to figure out what the addition of wings would have done to its weight and performance numbers.

The second addition to this launch system will be to fly the cargo spacecraft directly to the lower end of a 200-kilometer long basic non-rotating Skyhook as described in **Chapter 15**. This will reduce the speed required for orbit by another 260 meters per second. Adding these two values together gives a total speed reduction of 1,060 meters per second, which can now be subtracted from the speed of the second stage of the Titan 2.

The end result of these additions to the launch system is to reduce the propellant mass of the second stage by 2,450 kilograms and to increase the useful payload of the cargo spacecraft from 1,400 kilograms to 3,850 kilograms.

Now compare that to the useful payload of the Titan 2 when flying direct to a space station without the assistance of a ground accelerator or a Skyhook.

Useful payload to space station, no ground accelerator, no Skyhook 1,260 kg

Useful payload to space station, with ground accelerator, with Skyhook 3,850 kg

Notice that the useful payload has increased by a factor of three. This also means that that the cost of going to the space station with the Skyhook is one-third the cost of going to the space station without the ground accelerator and Skyhook.

To put this in terms of the cost to fly to a space station, Denis Tito paid $20 million to fly to the International Space Station in 2001. If the Soyuz launch vehicle had been ground accelerator launched to 600 MPH and the space station had been equipped with a 200-kilometer long basic non-rotating Skyhook, the cost for that ticket would have been $6.7 million.

While this cost reduction is not enough to make spaceflight mass-market affordable, it is still a significant improvement. It is also just the beginning of what is possible.

Example 2 As previously mentioned, one of the advantages of the Skyhook concept is its ability to gradually grow with the market and how the cost of going to orbit goes down with each increase in the length of the Skyhook cable. Therefore, this example will explore the impact of increasing the overall length of the basic Skyhook used in the first example from 200 kilometers to 380 kilometers. Everything else will remain the same. The launch vehicle and spacecraft will remain the same and the ground accelerator will remain the same. As before, reductions in propellant mass for the second stage will be added to the useful payload in the cargo spacecraft. Also as before, there will be no change to the size of the second stage due to reducing the amount of propellant it will carry, nor will there be any change to the size of the cargo spacecraft due to increasing its payload. It is assumed for these examples that any weight savings that would result from downsizing the second stage would be used to increase the size of the cargo spacecraft and the end result would be zero change in weight.

As previously shown, the 200-kilometer overall length basic Skyhook reduces the speed required for orbit by 260 meters per second and the 380-kilometer Skyhook reduces it by 505 meters per second. This additional reduction in speed required reduces the second stage propellant by an additional 696 kilograms, which is then added to the useful payload. The end result of this minor addition to the Skyhook is that the seat price for flying to the International Space Station is reduced from $6.7 million to $5.5 million.

Example 3 For this example assume a mature Skyhook with a 2,300 meters per second velocity reduction at the lower endpoint, a 600 MPH ground accelerator located on a 12,000-foot tall mountain with a velocity reduction of 800 meters per second, and a wingless rocket powered reusable first stage with expendable upper stage launch vehicle. Assume the first stage has an operating empty weight of 15% of the total first stage weight when operating as an expendable stage, 20% when operating as a reusable stage, and has a specific impulse of 320 seconds. Assume the expendable upper stage has an operating empty weight of 12% of the second stage weight and a specific impulse of 460 seconds. Also, assume that the mass of the delivered payload will be 35% of the gross payload mass of the launch vehicle and the size of the delivered payload will be 3,000 kilograms.

The numbers for this launch vehicle when operating as an expendable launch vehicle without the ground accelerator and Skyhook are:

1st Stage	empty	21,000 kg
	propellant	119,000 kg
	stage total	140,000 kg
	Isp	320 seconds
2nd Stage	empty	6,000 kg
	propellant	44,000 kg
	stage total	50,000 kg
	Isp	460 seconds
Gross payload to LEO		8,600 kg
Total		198,600 kg
1st Stage apparent velocity		2,867 m/s
2nd Stage apparent velocity		6,265 m/s
Total apparent velocity		9,132 m/s
Gross Payload Fraction		4.33%
Empty weight fraction		13.60%
Propellant fraction		82.07%
Useful payload fraction		1.51%

If it is assumed that this vehicle costs $60 million to fly as an expendable vehicle, then the cost per kilogram delivered to a space station is approximately $20,000 per kilogram.

Redesigning this vehicle for use with a 600 MPH ground accelerator, a reusable first stage, and a mature Skyhook gives the following numbers:

1st Stage	empty	8,800 kg
	propellant	35,200 kg
	stage total	44,000 kg
	Isp	320 seconds
2nd Stage	empty	1,800 kg
	propellant	13,200 kg
	stage total	15,000 kg
	Isp	460 seconds
Gross payload		8,600 kg
Total		67,600 kg
VALT apparent velocity		800 m/s
1st Stage apparent velocity		2,305 m/s
2nd Stage apparent velocity		3,695 m/s
Skyhook apparent velocity		2,300 m/s
Total apparent velocity		9,100 m/s
Gross Payload Fraction		12.72%
Empty weight fraction		15.68%
Propellant fraction		71.60%
Useful payload fraction		4.44%

Notice how much smaller the launch vehicle has become. Its take-off weight has gone from 198,600 kilograms without the ground accelerator and Skyhook to 67,600 kilograms with the ground accelerator and Skyhook. That is one-third the size.

Cost The baseline expendable launch vehicle without the ground accelerator and Skyhook was assumed to have a launch cost of $60 million. If the launch cost is divided by the empty weight of the first and second stages, the result is the cost per kilogram to build and fly the baseline vehicle. That comes out to $60M/27,000 kilograms or approximately $2,222 per kilogram to build and fly.

If the empty weight of the launch vehicle that is designed to fly on the ground accelerator and Skyhook is multiplied by this value it will give a ballpark estimate of what this vehicle would cost to build and fly as an expendable launch vehicle. That comes out to $23.6 million.

Since it is supposed to be a reusable first stage launch vehicle it will be appropriate to reduce this value by some amount to account for the savings that are made possible by recovering and reusing the first stage. This was covered more in **Chapter 13** where it was assumed that the reusable first stage would have an initial flight life of 10 flights, that it represented 75% of the cost of the launch vehicle, and that refurbishment costs per flight would be approximately 10% of the stage cost per flight. This led to a cost reduction of 60% per flight.

Subtracting 60% from $23.6M leaves a cost per launch of $9.4M.

Dividing that by the useful payload of 3,000 kilograms gives a launch cost of $3,140 per kilogram. That comes out to less than one-sixth the cost of the baseline expendable launch vehicle without the ground accelerator and Skyhook. Putting this in terms of seat cost and assuming 150 kilograms per passenger, the cost per person to the Skyhook using this launch system becomes $471,000.

Example 4 So just how low can the cost of flying to a Skyhook equipped space station go once the Skyhook has grown to full size and complete reusability of the launch vehicle and air-breathing propulsion are included? Pretty low.

Back in the 1990s when work was being done on the idea of a combination air-breathing and rocket powered reusable spaceplane that could fly all the way to orbit, the idea of using such a spaceplane to fly to the lower end of a Skyhook was explored. The end result of that study was a horizontal take-off spaceplane that had a take-off weight of approximately 165,000 kilograms that could deliver a 16,000 kilograms gross payload to the lower end of a mature Skyhook. The estimated cost to orbit for this system when operating with a flight rate of 72 flights per year was $1,200 per kilogram.

X-43 test vehicle, NASA artwork

Putting this in terms of seat cost again, if it is assumed that the useful payload for this vehicle when carrying passengers is 12,000 kilograms and the average passenger with baggage and supplies is 150 kilograms; the cost per seat will be approximately $270,000.

That comes out to a little more than a penny on the dollar compared to what Dennis Tito paid: and that is at a flight rate of one flight every five days. When it gets to the point that there are multiple flights per day the price will become much less. This is why the NASA representatives who first reviewed the suborbital spaceplane with non-rotating Skyhook concept said,

This is the first idea we have seen that offers a believable path to $100 per pound to orbit.

The Central Space Station in the Earth's shadow with its external lights on.
By Nick Stevens

Chapter 20: To the Moon, Mars, and Beyond

No matter what your dreams of space are; living on and exploring Mars, building space colonies and satellite solar power stations, or mining the asteroids, without affordable to the individual spaceflight, they will always remain dreams.

Commercially available, affordable to the private individual spaceflight has been the dream of just about everyone who has ever wanted to travel into space. Its existence has been one of the most basic assumptions of almost every science fiction novel and movie ever produced. It is also a fundamental technological requirement that will need to exist if space tourism, orbiting hotels, space colonies, satellite solar power stations, cities on the Moon and Mars, and asteroid mining are ever going to happen in the real world. It is that basic.

However, there is more to making all these dreams into reality than making the journey to Earth orbit affordable. In order to get to a space colony, to the Moon, to an asteroid, or to Mars, it will be necessary to accelerate to escape velocity and beyond, and that part of the journey will also need to be affordable.

The Journey to the Moon, Mars, the Asteroids and Beyond

When Wernher von Braun first proposed a manned expedition to Mars in 1948, he envisioned a fleet of 10 spacecraft carrying a crew of 70 people that was to be assembled in Earth orbit. These 10 spacecraft were to be powered by chemical rocket motors and were designed for four major propulsion events:

- to accelerate from Earth orbit onto a Mars transfer orbit;
- to decelerate from the Mar transfer orbit into Mars orbit;
- to accelerate from Mars orbit onto an Earth transfer orbit; and,
- to decelerate from the Earth transfer orbit into Earth orbit.

The propellant for each propulsion event was kept in separate tanks so that the empty tanks and their support structures could be jettisoned at the end of each event. The same rocket motors were used for the entire voyage. At the end of the journey, what arrived back in Earth orbit was only a tiny fraction of what left. It was a use once, single voyage fleet.

The travel time from Earth orbit to Mars orbit was 260 days. Total trip time, including a 449-day stay at Mars, was two years and eight months.

Some of the major unknowns regarding Dr. von Braun's proposed expedition to Mars were the possible negative effects of long-term zero gravity on the crew, and the health risks due to cosmic radiation and solar flares.

The total mass of this fleet at the moment of departure from Earth orbit was to be 37,200 metric tons or 531.4 metric tons per crewperson. The total amount of propellant carried by the fleet was 35,555.5 metric tons, which worked out to a propellant fraction of 95.58%. Breaking it down by the number of people making the journey, it came out to 507.9 metric tons of propellant and 23.5 metric tons of ship, supplies, and payload per person. All of which had to be lifted into Earth orbit on a fleet of chemical powered multi-stage rockets.

To give an idea of just how big an undertaking this would have been, the International Space Station has an orbital mass of approximately 420 metric tons, and it took more than 115 rocket launches using five different launch systems over a span of years to lift all the parts and pieces into orbit. For the Apollo program, the total tons per person for going from Earth orbit to the Moon and back was approximately 53.9 metric tons. Dr. von Braun's expedition to Mars was 10 times that on a per person basis. Obviously, the cost of this mission to Mars was also a very big issue.

Approaching Mars, 1948 *By Chesley Bonestell*

Ernst Stuhlinger In order to reduce the cost of a manned Mars expedition, the Director of NASA Marshall Space Flight Center's Space Sciences Laboratory, Ernst Stuhlinger, proposed three different ion powered spaceships for traveling to Mars between the years 1954 and 1962. The first of these concepts was solar powered while the second and third were nuclear powered. While not originally proposed as reusable vehicles, it is possible that all three of these designs could have been built as reusable spacecraft making them possible candidates for an ongoing transportation system between Earth and Mars. A fourth concept that he proposed in 1966, called the NERVA-Electric Manned Mars Vehicle, had an expendable nuclear thermal first stage and as a result could not be made into a reusable transportation system.

The Sun Ship The first of these designs, known as the Sun Ship (1954), had an Earth departure mass of 280 metric tons, a crew and cargo mass of 50 metric tons, and a solar electric power system mass of 167.2 metric tons. It was supposed to consume only 18.6 metric tons of propellant during one year of continuous thrusting. The number of crew, the empty weight of this vessel, its travel time, and what it used for a Mars lander are unknown.
For a short video showing the Sun Ship see reference 56.

Ernst Stuhlinger's solar electric Sun Ship approaching Jupiter.
By Graham Gazzard and Nick Stevens

The Umbrella Ship The second design, known as the Umbrella Ship (1957), had a nuclear electric ion propulsion system, a 20-person crew, and a departure mass of 660 metric tons. The total mass per crewperson was 33 metric tons.

This was a very big improvement over Dr. von Braun's original concept. The ship carried 332 metric tons of propellant and 136 metric tons of payload including a Mars lander. The remaining 192 metric tons was the empty weight of the ship.

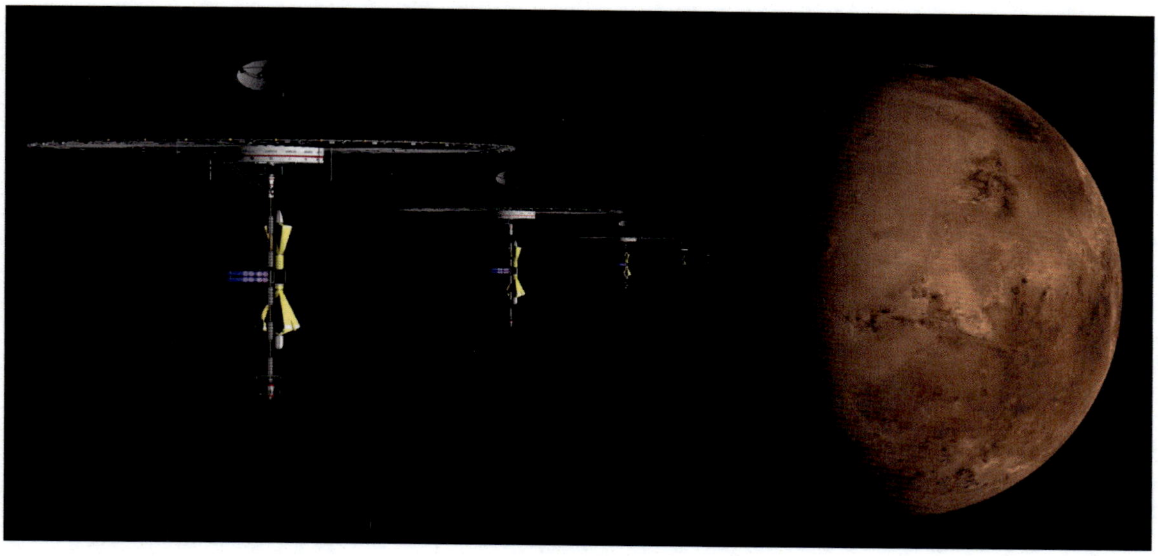

Ernst Stuhlinger's fleet of nuclear electric Mars Umbrella Ships approaching Mars. By Nick Stevens

---

If it is assumed that the Umbrella Ship was reusable and could be used for multiple voyages, it then becomes possible to calculate the recurring cost per crewperson for the voyage by subtracting the empty mass of the spaceship from the Earth departure mass and dividing the result by the number of people onboard. This particular measure of performance is a very useful term for comparing the relative costs of different Earth/Mars transportation concepts. For the Umbrella Ship, the total recurring mass per crewperson was 23.4 metric tons.

Travel time from Earth orbit to Mars orbit with this vessel was 402 days. Total trip time, including a 476-day stay at Mars and the return to Earth, was 1,276 days, or three years and six months. In order to resolve some the concerns about the possible negative effects of prolonged zero-gee, the crew quarters for this ship were a spinning wheel so that the crew had a small amount of artificial gravity for the duration of the journey.

A total of 10 ships were to make the trip in order to provide mutual support in the event of an emergency. However, the total number of ships could vary as each ship carried its own lander and all the supplies it needed for the journey.

The main disadvantages of this concept were the total time of the voyage, the cost of a 10-ship fleet, and concerns about radiation exposure from the Van Allen radiation belts around the Earth, cosmic radiation, and solar flares.

For a short video showing the Umbrella Ship see reference 57.

The Martian Electric Spaceship The third design, the Martian Electric Spaceship (1962), also known as the Ion Rocket, was the fastest of the three. This ship would take 56 days to accelerate out of low Earth orbit to Earth escape velocity, and then continue accelerating in solar orbit until it was half way to Mars where it would then turn around and begin decelerating into high Mars orbit. The transfer time in solar orbit, from Earth escape velocity to Mars escape velocity, was 146 days. After that, it would take an additional 21 days to spiral down to low Mars orbit. Total travel time for this spacecraft from Earth departure orbit to low Mars orbit was 223 days. The stay time at Mars for this mission concept was only 29 days. The return trip from low Mars orbit took 18 days to reach Mars escape velocity, 268 days for the transfer in solar orbit, and 31 days to slow from Earth escape velocity back down to its original departure orbit. Total voyage time from Earth orbit to Mars and back was 569 days, or one year and seven months.

Ernst Stuhlinger's Martian Electric Spaceship underway in Earth orbit on its way to Mars. By Nick Stevens

---

The planned fleet size for this expedition was five ships, three that would carry Mars landers, and two that would not. The crew size for each ship was three people. Departure mass for each ship was 360 metric tons. The ships with Mars landers would depart with 120 metric tons of propellant, the ships without Mars landers would carry 190 metric tons. The Mars landers were 70 metric tons each. As with all the other concepts, the Mars landers were single use expendable vehicles that were discarded in Mars orbit prior to the return to Earth.

Prior to the start of the return voyage, the remaining propellant on the five Earth/Mars spaceships was to be evenly distributed between them. The return to Earth departure mass was 170 metric tons per ship. In addition, each ship carried a 50-metric ton radiation shelter and the ships were spun so that the crew experienced 1/10 Earth normal gravity for the duration of the voyage.

The total mass per crewperson for this concept is 120 metric tons. If it is assumed that these vessels were reusable the recurring mass per crewperson is approximately 65 metric tons. The much higher tonnage per crewperson compared to the Umbrella Ship, both the total and the recurring, is due to the mass of the radiation shelter, which also takes away from the size of the crew. *For a short video showing the Mars Ion Rocket see reference 62.*

Comparing the Concepts The most important item to review when comparing these and other space transportation concepts is the recurring tons per crewperson that needs to be lifted from the surface of the Earth into low Earth orbit for each concept. This figure includes the amount of food, air, water, propellant, and any other non-reusable equipment that would be consumed per person for the trip. The lower this value is, the lower the recurring cost of going to the Moon, Mars, an asteroid, or anywhere else will be.

This can be seen very clearly by comparing the recurring mass figures for the concepts that have been examined so far.

- Dr. von Braun's original Mars concept 531.4 MT/crew member
- the Apollo program, Moon 53.9 MT/crew member
- the Umbrella Ship, Mars 23.4 MT/crew member
- the Martian Electric Spaceship, Mars 65.0 MT/crew member.

It is also important to keep in mind that as good as either the Umbrella Ship or the Martian Electric Spaceship looks in comparison to von Braun's original concept; neither of them were built due to their high cost. Costs that were driven by the one time cost of design, development, construction, and launch, but also by the recurring cost per person for follow-on missions.

A good example of this can be seen by using the recurring mass figure for the Martian Electric Spaceship. This is probably the most realistic of the three concepts due to the 50-ton radiation shelter that was included in the design. If it is assumed that the launch cost using a late 1960s era expendable launch vehicle was $2,000 per kilogram, the cost of launching all the recurring mass into Earth orbit is $130 million per person. Obviously not a price that very many people could afford, then or now. Even if the cost of getting to Earth orbit could be gotten down to $250 per kilogram, the recurring cost per person for going to Mars using the Martian Electric Spaceship becomes $16.25 million - still not mass-market affordable.

Why Mars?

We are all . . . children of this universe. Not just Earth, or Mars, or this system, but the whole grand fireworks. And if we are interested in Mars at all, it is only because we wonder over our past and worry terribly about our possible future.

Ray Bradbury, 1973, from 'Mars and the Mind of Man'

Hubble Space Telescope

There are celestial sights more dazzling,
spectacles that inspire more awe,
but to the thoughtful observer who is privileged to see them well,
there is nothing in the sky so profoundly impressive
as the canals of Mars.
Fine lines and little gossamer filaments only,
cobwebbing the face of the Martian disk,
but threads to draw one's mind after them
across the millions of miles of intervening void.

Percival Lowell, 1908

NASA photo

Ten thousand times a hundred thousand dusty years ago,
Where now extends the Plain of Gold,
Did once my river flow;
It stroked the stones and spoke in tongues and splashed against my face,
Till ages rolled, and the Sun shone cold,
On this unholy place.

Jonathan Eberhart, 'Lament for the Red Planet'

NASA photo

Reducing the Cost of Going to Mars

There have been many ideas proposed for reducing the cost of going to Mars. Ideas that can significantly reduce the amount of propellant, supplies, and equipment that needs to be lifted into Earth orbit for a trip to Mars. Some of the better known ones are:

- sending all the equipment and supplies in advance of the crew on slow unmanned one-way spacecraft so that what would have been shielding mass for the crew, return propellant mass, and additional propellant for reducing the travel time becomes additional payload mass;
- making the Mars landers as single stage vehicles that are both reusable and refuelable using Martian resources;
- using Martian resources to made the propellant for the return trip to Earth;
- using either lunar resources or relocated asteroids for making the propellant for going to Mars; and,
- using a Mars Cycler spacecraft that travels between Earth and Mars and the asteroid belt on a cyclic orbit, where the passengers use small spacecraft to jump on and off the Mars cycler spacecraft as it swings by Earth, Mars, and the asteroids.

There are other less well-known ideas.

A Non-rotating Skyhook Transportation System for Mars

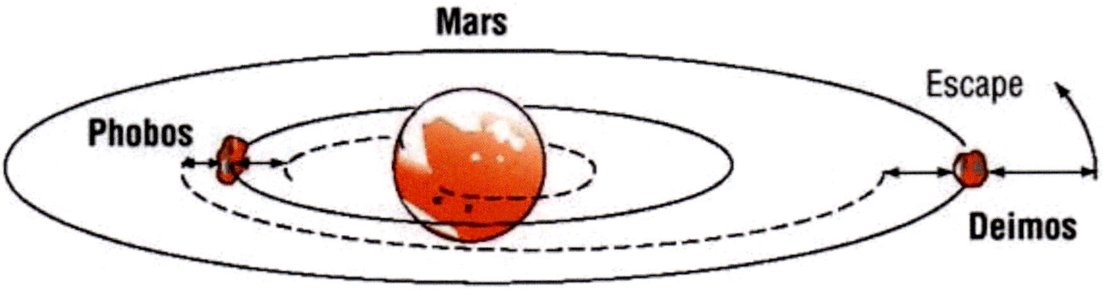

NASA artwork

In 1984, Paul Penzo of JPL proposed a planetary surface to escape velocity tether transportation system for Mars that consists of two unpowered non-rotating Skyhooks; one attached to the Martian moon Phobos, and the other attached to the Martian moon Deimos.

With this system, a spacecraft arriving at Mars, either direct from a Skyhook in orbit around the Earth, from an outpost space station at the Earth-Moon L2 Lagrange point, or from an Earth-Mars cycler spacecraft as it swings by Mars, docks at the upper end of the non-rotating Skyhook attached to the outer moon Deimos. The people and cargo on that spacecraft then transfer to an elevator that will take them down to the lower end of the Deimos Skyhook. There they board a small reusable Orbital Transfer Vehicle that will take them to the upper end of the non-rotating Skyhook that is attached to the inner moon Phobos. Again they transfer to an elevator that will take them to the lower end of the Phobos Skyhook where they will then transfer to the reusable single stage suborbital Mars Lander that will carry them down to the Martian surface. Passengers and cargo on the Martian surface that are bound either for the asteroid belt, or for Earth, would ride the system in reverse.

One of the advantages of this concept is that neither Martian Skyhook will require a propulsion system for orbital reboost and control, as they both will use the mass of the Martian moon they are attached to as a momentum bank to make up for any differences in the upward versus downward mass flow of people and cargo. This elimination of the propulsion systems for the two Skyhooks also makes for a significant reduction in the cost to build, and the cost to operate. Like the non-rotating Skyhook for Earth, this two-stage non-rotating Skyhook system for Mars can be affordably built with existing materials and technology.

The advantage of this system and how it reduces the cost of going to Mars is two-fold. For people arriving from Earth it eliminates the propellant for slowing down from Mars Transfer Orbit velocity to Low Mars Orbit velocity. In addition, since the Earth-Mars spacecraft does not have to slow down to Mars orbit velocity it also reduces the transit time between Earth and Mars, which reduces the amount of supplies they will need to carry.

This reduced transit time, reduced supplies requirement, and reduced propellant requirement also applies to people leaving Mars. Whether they are returning to Earth or heading out to the asteroid belt, the Martian Skyhook system eliminates the propellant that would have been used accelerating from low Mars orbit velocity to Mars escape velocity and beyond, which reduces the transit time, and the amount of supplies, all of which reduces the cost. The Martian Skyhook system also reduces the amount of propellant needed for getting from the Martian surface to Mars orbit.

Outpost Space Stations

In 1984, Paul Keaton, of the Los Alamos National Laboratory, wrote a paper about the need for a system of outposts in space along the routes to the planets and other places of interest. These outpost space stations to be warehouses for propellant, spare parts and supplies, rest stations for passengers and crew, and transshipment sites for cargo. The first outpost space station in this system is to be a space station located in low Earth orbit.

For the second outpost space station he proposed two possible locations, the Sun-Earth L1 Lagrange point, or the Earth-Moon L2 Lagrange point. Both of these locations are relatively easy to reach, and both allow the use of gravity assist maneuvers by departing spacecraft as they pass by the Moon and/or the Earth on their way to a near-Earth asteroid, Mars, Ceres, the asteroid belt, the next outpost space station, or beyond.

He also proposed that a third outpost space station be located at the Sun-Mars L1 Lagrange point where it would serve as the departure point for trips to the Martian moons, low Mars orbit, the asteroid belt, and return trips to Outpost Space Station 2. Since these outpost space stations would be located on the Interplanetary Transport Network they would also support the low energy transfer of equipment and supplies via unmanned spacecraft that could be sent ahead of a manned mission.

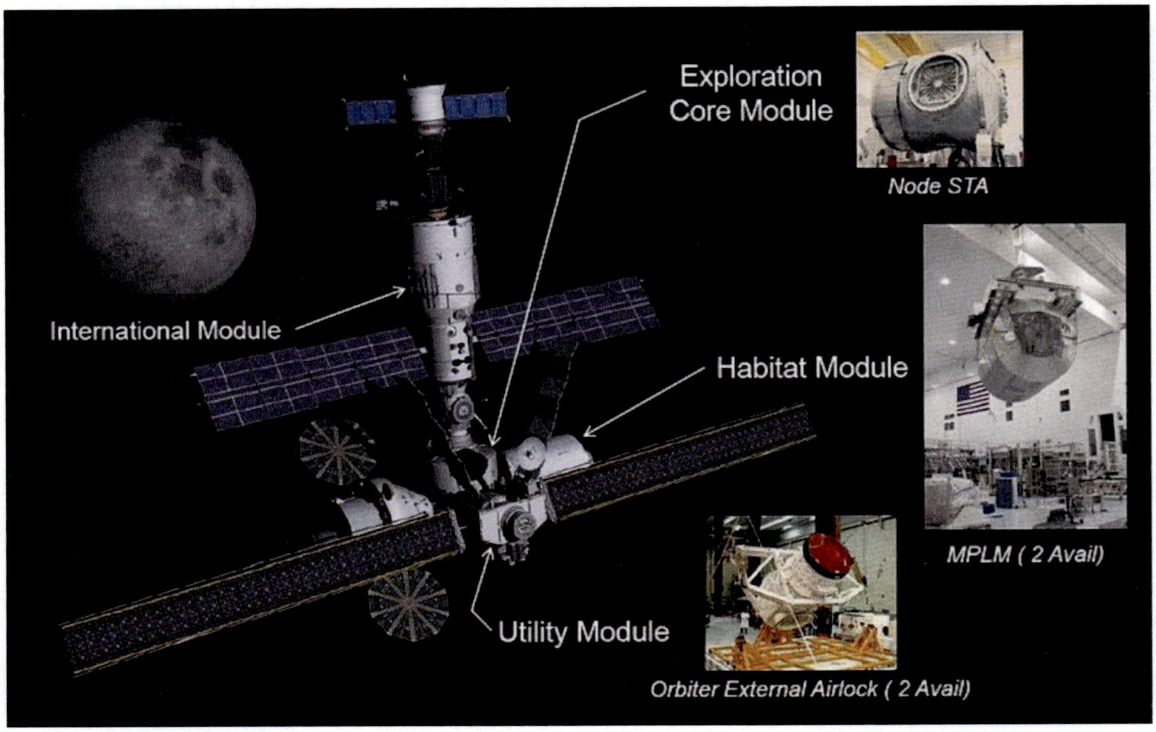

A possible first generation outpost station at the Earth-Moon L2 Lagrange point, NASA image

---

Another one of the advantages of the outpost space station concept is the way it breaks up the journey to Mars into segments. For example, once both the second and third outpost space stations are in place and local sources of propellant are established for each, Mars-bound spacecraft that travel between these two stations would only need to carry enough propellant for a one-way trip between the stations and not for the entire round trip journey to Mars as they would without the outpost space stations. This allows the spacecraft to carry a much larger payload fraction and to be designed as a completely reusable vehicle, two features that will greatly reduce the cost of interplanetary spaceflight.

An outpost space station with propellant depot, NASA artwork

This outpost space station system will also allow different types of spacecraft to be used for each leg of the journey so as to minimize both the overall travel time and the overall cost. For example, an Orion spacecraft could be used for travel between the upper end of an Earth orbiting Skyhook and Outpost Station 2 at the Earth-Moon L2 Lagrange point. Travel time for this spacecraft on this leg would be only six days as compared to the 56 days it would take Ernst Stuhlinger's Martian Electric Spaceship to go from low Earth orbit to escape velocity.

For the journey between Outpost Station 2 and low Mars orbit, a spacecraft along the lines of Nautilus-X Extended Duration Explorer could be used. Travel time from Outpost Station 2 to low Mars orbit using this spacecraft with an ion propulsion system would be approximately 167 days, the same as the Martian Electric Spaceship after it reached Earth escape velocity.

For a short video showing the Nautilus-X spacecraft see reference 60.

Paul Keaton published his paper on outpost space stations in 1984, the same year that Paul Penzo published his paper on the dual non-rotating Skyhook transportation system for Mars. Considering the timing, it is most likely that neither of these gentlemen knew about the other's work at the time they published. This is unfortunate as it is likely that the dual non-rotating Skyhooks attached to the Martian moons would make a better choice for where to locate Outpost Station 3.

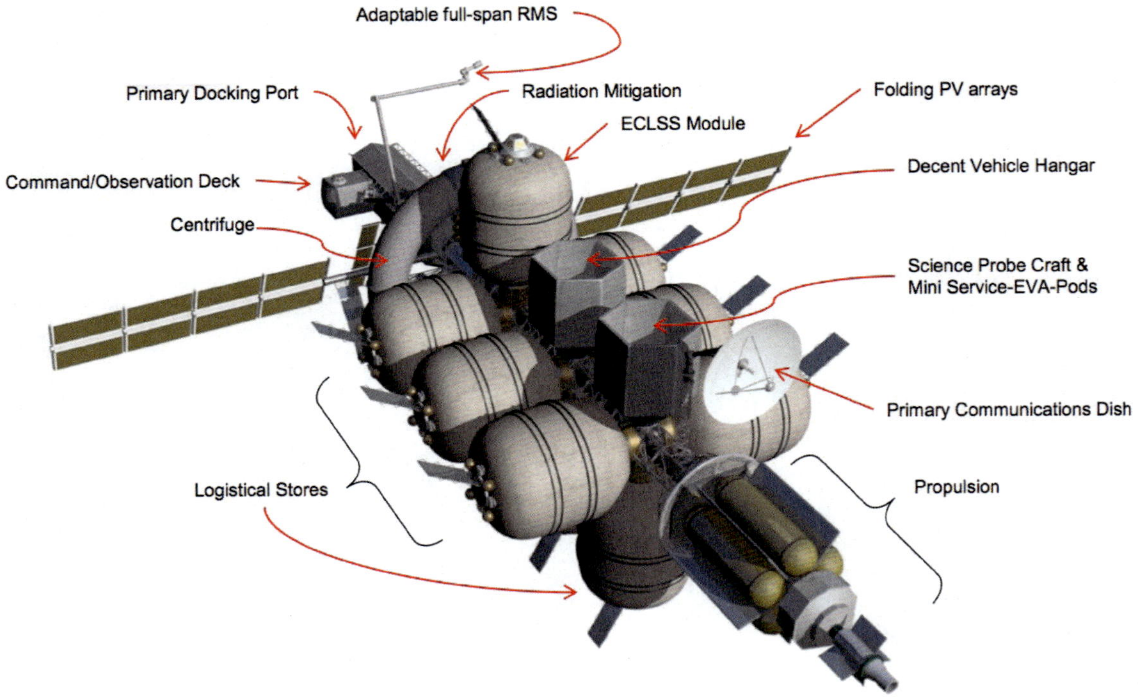

Nautilus-X Extended Duration Explorer, NASA artwork

A third possible candidate for Outpost Station 3 would be an ion powered non-rotating Skyhook for Mars similar to the one in Earth orbit. The advantage of this is that it could be built and tested in Earth orbit prior to sending it to Mars using its onboard ion propulsion system.

The advantage of either of these Martian Skyhook systems would be to allow the Nautilus-X spacecraft to fly by Mars on a cycler orbit. Not needing to slow down to Mars escape velocity or Mars orbit velocity, would significantly reduce the transit time between Outpost Station 2 and Mars, as well as reduce the amount of propellant required by Nautilus-X for the voyage. Transfer from Nautilus-X to the upper end of the Martian Skyhook would be performed using one of the Orion spacecraft that had docked with Nautilus-X at Outpost Station 2. The Mars arriving Orion spacecraft would use its onboard propulsion system to adjust for any difference in velocity between itself and the upper end of the Martian Skyhook.

Note It is important to realize that a third outpost station is not required for a manned expedition to Mars as it is possible to go directly from Outpost Station 2 to low Mars orbit. However, a third outpost station in Mars orbit that includes both a non-rotating Skyhook and a propellant depot would be a cost-effective addition to a permanent Earth-Mars transportation network should it be decided to make Mars into a second planetary home for mankind.

Another option that would reduce the propellant needs of a Mars arriving Orion spacecraft would be to equip the upper end of the Martian Skyhook with an additional tether-on-a-reel system that could be extended and used to capture a spacecraft moving in excess of Mars escape velocity as it flew by. Once the Orion spacecraft is firmly attached to the outer end of this additional tether-on-a-reel system, the tether would reel the Orion spacecraft into the upper endpoint station. Using this additional tether/reel system in reverse, it could also be used to launch a departing Orion spacecraft at speeds in excess of Mars escape velocity so that it would be able to match speed with a passing Mars cycler spacecraft without using any of its onboard propellant.

As with so many proposed space transportation concepts, the question usually becomes, "How real is it?" For the outpost space station concept, it is interesting to note that NASA has funded three different design studies for the building of such an outpost at the Earth-Moon L2 Lagrange point called Deep Space Habitat, Exploration Gateway Platform, and Skylab II.

Local sources of propellant

A key ingredient that the outpost space stations will need to supply if they are to make interplanetary spaceflight mass-market affordable will be local sources of propellant. NASA has been working on a possible solution to this with its proposed Asteroid Redirect Mission. This is the mission that will move a small asteroid, or a small piece of a larger asteroid, into lunar orbit. This asteroid redirect technology would allow small, high water content asteroids to be moved to Outpost Station 2 at the Earth-Moon L2 Lagrange point, or to Outpost Station 3, where they would be processed for making drinkable water, oxygen for breathing, and rocket propellant. Another possible source of local propellant for Outpost Station 2 is the Moon itself. Ice deposits have been detected in the bottoms of permanently shaded craters located near the north and south poles of the Moon. With affordable access to the lunar surface using a single stage reusable lunar lander, this becomes a second possible source.

A third possible method for accessing local sources of propellant is the Kuck Mosquito. This idea would use an unmanned spacecraft equipped with a drill, a pipe with a heating element, and a storage tank, to extract water and other frozen volatiles from an asteroid. Once the storage tank is full, the spacecraft would transport the volatiles to an outpost space station for processing.

The most important thing about a local source of propellant near the Moon is that it could be used by the Orion spacecraft as it travels back and forth from the upper end of the Earth orbiting Skyhook to Outpost Station 2, to lunar orbit, and out to a near-Earth asteroid. This local source of propellant could also be used by Nautilus-X on its voyages of exploration to Mars and around the inner solar system once it is built and flying. Both of these applications will noticeably increase the payload fractions of both spacecraft, will dramatically reduce the amount of propellant that needs to be brought up from the Earth, and will significantly reduce the cost of flying these missions.

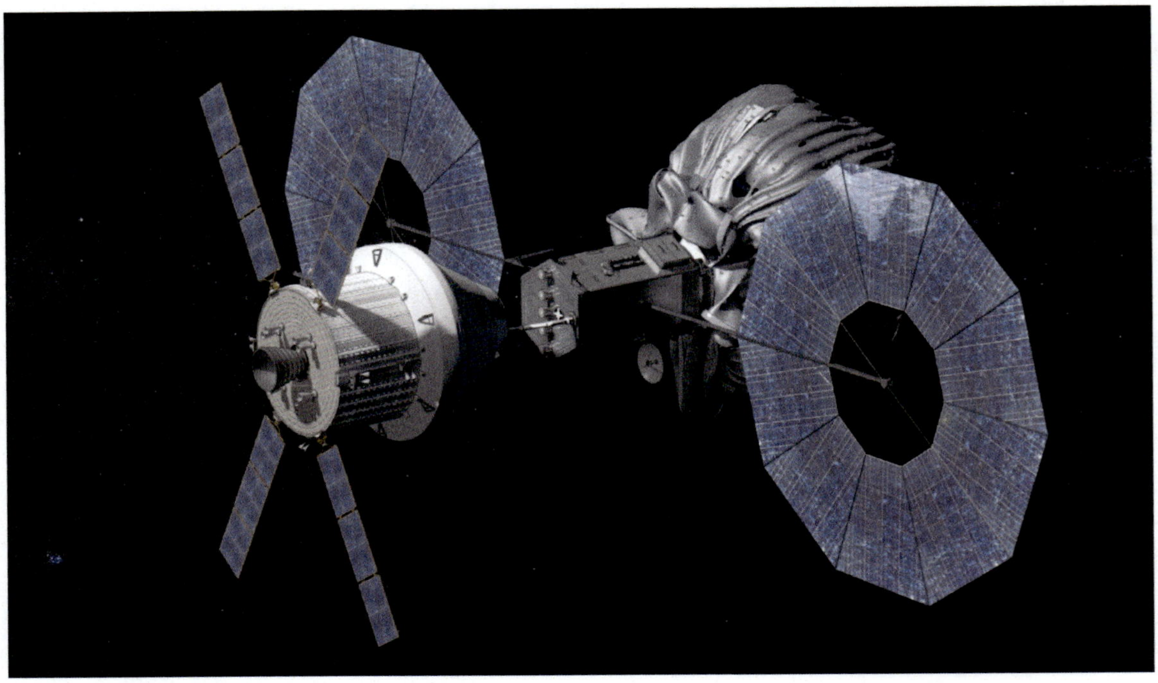

Orion docking with the captured asteroid in lunar orbit, NASA artwork

The Kuck Mosquito, from the paper 'The Deimos Water Company' by David L. Kuck. Graphics by Nick Stevens

Combination Systems

There are many ways to combine these ideas in order to build an affordable transportation system that will allow us to explore, develop, and settle the solar system. As already stated, the best combination will probably change over time as the market grows and as the space economy expands and develops. It will also depend on what we choose to give priority to. Will we choose to develop orbital tourism, orbiting hotels, asteroid mining and orbital industries to start with, or will we choose to give priority to a manned expedition to Mars? Will we choose to build space colonies, or will we choose the settlement and terraforming of Mars in order to build a new planetary home for mankind? The number of choices and the potential directions that we can choose are really pretty amazing once spaceflight becomes truly affordable.

Mars

One possible direction that might be chosen will be to send a manned expedition to Mars. If we are going to Mars both for science and as a precursor for settlement, we will need a space transportation system that includes a non-rotating Skyhook in Earth orbit with an upper endpoint velocity that is close to Earth escape velocity. We will also need an Outpost Space Station at the Earth-Moon L2 Lagrange Point with a local source of propellant and the Orion spacecraft for moving astronauts from the upper end of the Skyhook to lunar orbit and the Outpost Space Station at L2.

In preparation for the first manned expedition to Mars, a large volume of equipment and supplies will need to be sent on ahead and parked in Mars orbit. Some of these supplies will be delivered using an unmanned ion powered spacecraft that will shuttle back and forth between Outpost Space Station 2 and Mars orbit using a low energy transfer orbit, and some will be delivered by unmanned solar sail powered spacecraft on a one-way mission. Depending on our level of commitment to the long-term settlement of Mars, the equipment and supplies that are sent ahead will need to include such things as Mars habitats, Mars rovers, reusable single stage Mars Landers, inflatable greenhouses for growing food, packaged food and spare parts, propellant and propellant making equipment, orbital habitats, and possibly even cables and elevators for building a dual Skyhook system that will be anchored to the two Martian moons, Phobos, and Deimos.

The first manned expedition to Mars will most likely depart Outpost Space Station 2 in a Nautilus-X spacecraft. This spacecraft will only carry enough propellant and supplies for the trip to Mars, as the return propellant and everything else, will already be there. Keeping the spacecraft light will make possible a shorter travel time to Mars, which will keep the radiation exposure of the crew to a minimum.

Follow-on manned missions to Mars will also use a Nautilus-X spacecraft that departs from Outpost Space Station 2, but these spacecraft will most likely be placed on an Earth-Mars cycler orbit. This will allow for an even faster Earth-Mars transit and less radiation exposure. The actual arrival at Mars will be performed using an Orion spacecraft that is carried on the Nautilus-X spacecraft just for this purpose. The Orion spacecraft will release from the Nautilus-X as it swings by Mars and perform either an aerobrake reentry into the Martian atmosphere in order to slow down to Mars orbital velocity, or dock with the outer end of the Deimos Skyhook if that has been built.

Cost As with any large-scale operation, the total start-up costs for this Mars Transportation System will be high. The cost of the Nautilus-X spacecraft, Outpost Station 2, and the initial infrastructure equipment at Mars will not be cheap. But once the full transportation system is in place; the Skyhook at Earth, the Orion spacecraft, the Earth-Moon L2 Outpost Space Station with local propellant, the Nautilus-X spacecraft, the Skyhook system at Mars, the reusable Mars Landers with local propellant, and the locally grown food, the amount of use once hardware and consumables that will need to be launched from Earth per crewperson will drop tremendously. It is also important to remember that the cost for all this will be spread out over decades and that it is not an all at once expense. In truth, we have already started paying for it with the development of the Orion spacecraft, the work that is currently going on to develop and build an outpost space station, the ion propulsion systems used on Deep Space 1 and the Dawn spacecraft, the International Space Station, plus all the work that is going on in the commercial sector to develop low-cost reusable first stage launch vehicles, carrier aircraft, commercial space stations, etc., it all applies to this Mars Transportation System. In many ways it is like the building of the first transcontinental railroad, it was a major expense spread out over a number of years whose purpose was to make traveling back and forth across the country affordable. It also started earning money even before it was finished. That is how it will happen with this.

Recurring Cost So what will this Mars Transportation System do to the recurring tons per person that will need to be launched into Earth orbit for a trip to Mars? If it is assumed that a round trip tour of duty to Mars will last 3 years and that all the food and clothing for that person will need to come from Earth in the beginning, then the mass of recurring consumables per crewperson plus a prorated share of the spare parts will probably come out to somewhere in the neighborhood of 4 metric tons per person. Once local sources of food are developed that figure should drop to 3 metric tons or less.
Now compare those numbers with the earlier concepts.

- Dr. von Braun's original Mars concept — 531.4 MT/crew member
- the Apollo program, Moon — 53.9 MT/crew member
- the Martian Electric Spaceship, Mars — 65.0 MT/crew member
- this Mars Transportation System — 3 to 4 MT/crew member.

Space Tourism, Orbital Hotels, and Orbital Factories

Another possible direction that might be chosen for our next step in becoming a spacefaring people is the development of orbital tourism and orbital industries. For this to happen the cost of getting into Earth orbit needs to be as low as possible. That means growing the Skyhook as quickly as possible and developing the lowest possible cost launch vehicle for flying to the bottom of the Skyhook. As the cost of going to orbit goes down and the number of people in orbit increases, finding a space-based source of water and oxygen for those people will become a commercially worthwhile proposition. That leads to a couple of possible choices; redirect a small high water content asteroid either into lunar orbit, to one of the Earth-Moon Lagrange points, into high Earth orbit, or into low Earth orbit. Another possible choice is a return to the Moon for the water ice that has been found in the permanently shaded regions at the lunar poles. Of the two, the redirection of a small high water content asteroid is probably the most cost-effective solution. The only remaining issue is where to put it. While NASA has proposed lunar orbit for its first asteroid redirect mission, a high Earth orbit just a little further out than geosynchronous orbit might make another good choice. A low Earth orbit that is a little higher up than the upper end of the Skyhook is a third possible choice, but most likely moving even a small asteroid that close to Earth would probably make a lot of people a little uncomfortable.

NASA photo

Both

There is actually a very good reason that both of these directions should be pursued at the same time. NASA has stated that it is turning over Earth to orbit transportation to private industry and that it is planning to move outward to an Outpost Space Station at L2, to the near-Earth asteroids, and eventually to Mars. If private industry is building and developing the infrastructure for orbital tourism and orbital industries in the form of low-cost Earth to orbit transportation and low-cost Earth orbiting space stations, then the development of a NASA funded Mars project at the same time would speed up the growth of the Skyhook and the improvement of the launch vehicles for flying to it, that much faster. So in a strange delightful kind of way, the building of a transportation system for going to Mars and the development of space tourism and orbital industries could be very synergistic and beneficial to both programs.

Another possible synergy that could come from both projects going at the same time would be for NASA to turn over the relocation of small asteroids to commercial companies once they have proven the viability of the technology. With both NASA and orbital industries as customers for asteroid supplied water, propellant, and shielding, there would most likely be a number of companies competing to meet that demand. Turning over the development of these resources to private industry would save NASA money and would allow NASA to focus on what it does best: the exploration of our solar system.

A historical example of a similar situation is the U.S. government funding the building of the first transcontinental railroad back in the 1860s. Look at the explosion of settlement, development, jobs, and new industries that were stimulated by that project. Also, look at how it changed both the United States and the world for the better as large railroad systems were built all around the planet as a result of that project.

It is starting to sound like a very exciting future.

NASA artwork

Chapter 21: Space Colonies and Starships

It is difficult to say what is impossible, for the dream of yesterday is the hope of today and the reality of tomorrow.

Robert Goddard

In 1977, Dr. Gerard O'Neill, a Professor of Physics at Princeton University, published a book entitled *The High Frontier: Human Colonies in Space*. It is this book that brought the idea of large Space Colonies with Earth-like environments to the attention of the general public. His plan was to establish a self-sustaining, industrialized community in space where it would have a continuous view of the sun for generating solar power. The first Space Colony was to be built at L5, the trailing Lagrange Point in the Earth-Moon system, using lunar materials that would be launched from the surface of the Moon using a linear accelerator that he called a mass driver. The economic justification for this project was the building of large space-based satellite solar power stations that would supply the Earth with all the clean, affordable, electric power it could use. Electric power that would be beamed down to the Earth with low-density microwave beams at a price that was lower than that produced by conventional oil-fired power plants.

One of the things that made this idea so great was how it solved the problem of clean, unlimited, affordable power, and allowed us to move into space all at the same time. Another great feature of this concept was how all of this was to be done with the existing technology of the mid-1970s. It was a beautiful plan. All of it was within the scope of existing technology and all of it was economically workable based on the NASA advertised launch cost for the Space Shuttle of $50 to $100 per pound to low Earth orbit. Unfortunately, the advertised launch cost for the Shuttle was not even remotely close to what it turned out to be in reality. As a result, Space Colonies and space-based satellite solar power stations got put on the shelf until affordable Earth to orbit launch costs become a reality.

There are other aspects to Dr. O'Neill's work that did not get put on the shelf. For the first time, a physics professor had put forth a realistic vision of our future as a spacefaring species that was based on existing technology and sound engineering. And what a vision it was; thousands of man-made worlds floating in space that are 20 miles long and four miles in diameter, with over 120 square miles of usable land in each one, with forests and meadows, rivers and lakes, trees and animals, clouds and real weather, scattered throughout the solar system. What an incredible difference from all the science fiction dreams of the 1950s and 1960s of crowded little spaceships traveling between an overcrowded Earth and crowded little underground cities on the Moon, Mars, and in the asteroid belt. It is a vision of the future that makes all the previous dreams and visions incredibly small by comparison.

It is a vision that gives reality to these lines from Tennyson's Locksley Hall:

Many a night from yonder ivied casement, ere I went to rest,
Did I look on great Orion sloping slowly to the West.

Many a night I saw the Pleiades, rising thro' the mellow shade,
Glitter like a swarm of fireflies tangled in a silver braid.

Here about the beach I wander'd, nourishing a youth sublime
With the fairy tales of science, and the long result of Time;

When the centuries behind me like a fruitful land reposed;
When I clung to all the present for the promise that it closed:

When I dipt into the future, far as human eye could see,
Saw the Vision of the world, and all the wonder that would be;

Saw the heavens fill with commerce, argosies of magic sails,
Pilots of the purple twilight dropping down with costly bales.

Nor does Dr. O'Neill's vision stop here. With space colonies throughout the solar system, even as far out as Pluto, he also foresaw using the abundant solar energy that is available in space to gradually make enough antimatter that a space colony could embark on a journey to another star. It would not matter if there were Earth-like planets at the end of their journey as all they would need to survive would be comets and asteroids and the sunlight from their new star.

It is a vision that makes real the statement by Konstantin Tsiolkovsky where he says,

The finer part of mankind will, in all likelihood, never perish – they will migrate from sun to sun as they go out. And so there is no end to life, to intellect, and the perfection of humanity. Its progress is everlasting.

It is amazing when you stop and think about what we might become as star travelers, as individuals, as a civilization, and as a species. All the problems and troubles that we see in the world today are so incredibly small and meaningless when compared to what is possible. We truly live in a universe of unlimited possibilities, and all we have to do is reach out, to look beyond what has been done before, and find the courage to grow.

Chapter 22: Envisioning the Future

To create a new and better future you have to envision it first.
You cannot find this vision - you must create it.
You create it deep within you, with your heart, your soul, your education, and your brain.
Only then can you start to make it real in the world.
And once you have done that, you will have changed the world.

It is our vision of life and of the world that shapes everything around us. Some of our visions are very consciously chosen; some are based on subconscious or learned visions that have not been consciously examined. Some of our visions are based on a combination of the two. It is our individual visions that shape our personal lives. It is our group visions that shape our culture and the direction of our civilization.

You can see it everywhere around you. You can see it in how people select their significant other. Some people marry for romantic love. Some marry for physical beauty. Some marry for passion. Some marry for money and power. Some choose their mate based on their ability to dominate and control the relationship. Some people are looking to rescue or be rescued by someone. Some people commit fully to their relationship. Others act like they are ready to run at the first sign of any disagreement. The list goes on and on. Which one a person chooses depends on their vision of what a relationship is. The success or lack of success of a relationship inevitably rests on that vision.

This also applies to how people choose to view life and how they choose to make a living. Are they employees, are they self-employed, do they employ others? Do they do creative work or repetitive work? Are they social or asocial? Do they have a positive view of life, or a negative view? Is their cup half full, half empty, or is it running over? The lives people choose to live, the jobs they choose, and the joy or lack of joy they get from them, their success or lack of success, is all due to their vision of themselves, of life, and how they see themselves interacting with the world.

It also shows in their vision of the future. Do they envision a future that is better or worse than the world they see today? Do they see a Malthusian future with an ever increasing population, a declining standard of living, a collapsing ecology, and a collapsing civilization? Or do they see a better future where people use their brains, their education, their imagination and their creativity to envision and then build that better future?

The negative view of the future appears to be increasing. Polls have shown that the number of people who think that their childrens' lives will be worse than their own has become the majority viewpoint.

As Sally Ride once said, "Without a clear vision of the future, we flounder in the present." Based on the growing negative view of the world and our future, there appears to be a lot of floundering going on.

When this occurred in the past, a certain percentage of people picked up and moved on to greener pastures. It was this process that drove early modern humans to leave Africa and to gradually cover the globe. By their moving on, they also made things better for those who stayed behind.

In the beginning, moving on was easy. You picked up your possessions and you started walking. Gradually, over time, as the world became more crowded and the distances greater, moving on became more difficult and it required horses, then wagons, then ships, then railroads. Now it requires carrier aircraft, ground accelerators, reusable air-breathing rockets, and Skyhooks.

One thing hasn't changed and that is the need for a vision of a better life, and a better future. A vision that provides a direction, a focus for effort, and a path that leads to more abundance, more freedom, and an easier life with more opportunities. After all, without such a vision what is the point of moving?

When Europeans first started exploring and moving to North America, it was either to make a fortune, for land, or the opportunity to be their own person. Some wanted to be farmers, some to be tradespeople and business people, others still for religious freedom. All of them went for the opportunity to make their dreams and visions into reality. Another part of what drove them was the lack of opportunity at home. Europe at that time was experiencing population pressures that were at the limit of what could be supported with the technology of the day. As a result, society was becoming more and more stratified as those at the top sought to protect themselves and their lifestyles. This protective stratification process only served to further limit the available opportunities to everyone else. As a result, going to the New World became the only realistic option that was available to most people who had dreams of a better life. The New World had land available to anyone who was willing to clear it and farm it. In the New World, there were new villages, towns, and cities that were in need of craftsmen, tradespeople, merchants, anyone who was willing to work and who had a trade or a skill that was needed. The New World was a place where a person who was willing to work could go and build a new and better life. Yes, there was risk, but there was risk if they stayed home. At least in the New World, there was opportunity to go with the risk.

So that was the vision that drove them. It is the same vision that has driven every mass migration throughout the history of mankind. It is the vision of having the opportunity to make your dreams for a better life into reality.

In his book *The High Frontier: Human Colonies in Space*, Dr. O'Neill gave us a scientifically valid vision for a wonderfully abundant life that is full of opportunity as a spacefaring people. All that was missing from that vision was a scientifically valid concept for a mass-market affordable transportation system that can be built with existing technology that can take us to that new frontier. Ground accelerator launched or air-launched reusable air-breathing rockets for flying to a non-rotating Skyhook are a scientifically valid mass-market affordable transportation system that can be built with existing technology that will make Dr. O'Neill's vision into reality. It is a transportation system that will also make space tourism, orbital manufacturing, asteroid mining, and the manned exploration of Mars possible. It is a transportation system that will give us the solar system and someday the stars.

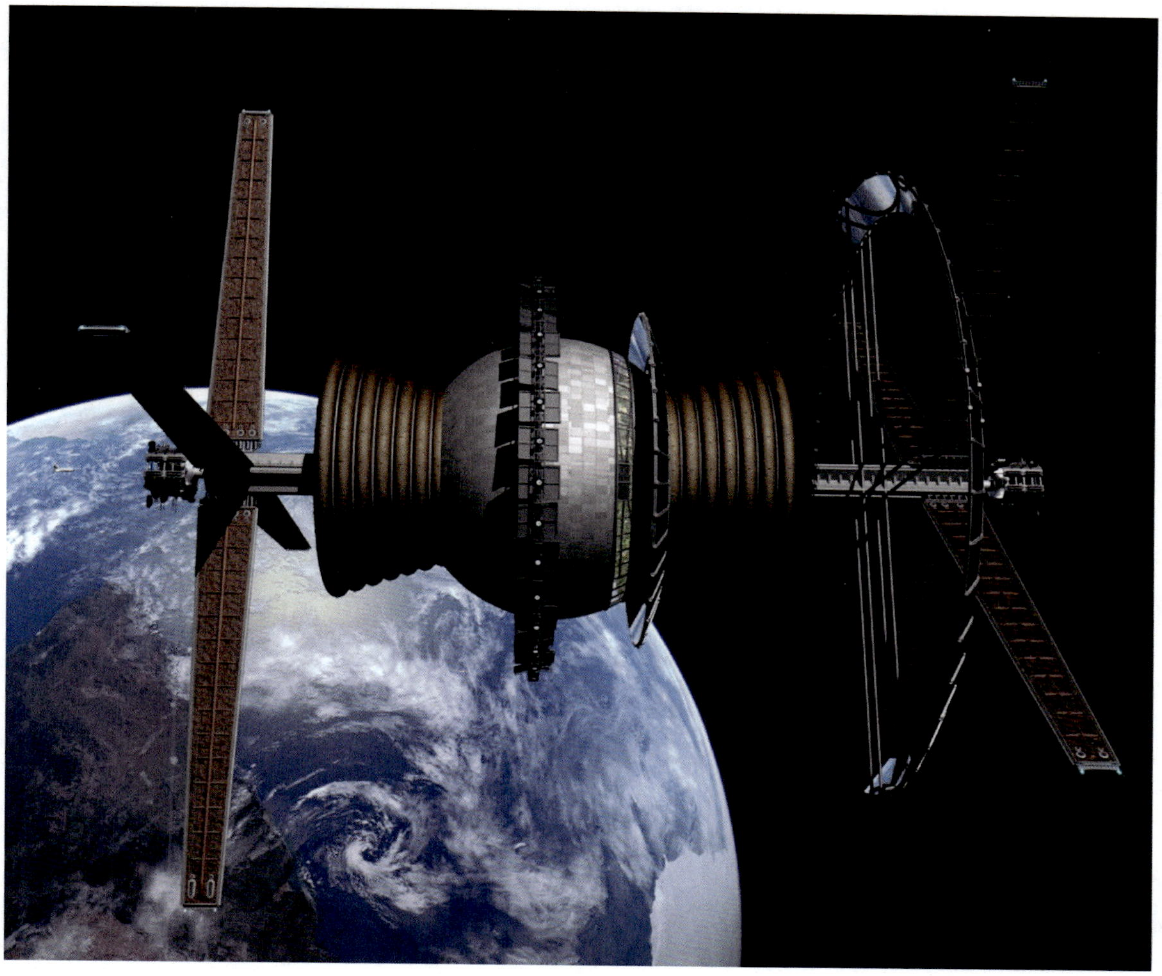

With affordable to everyone spaceflight, all the dreams of space colonization and space settlement become possible. This picture by Nick Stevens shows a space colony with an ion propulsion system approaching Earth on a gravity swing-by maneuver that will help boost its speed to escape velocity and send it on its way to Mars. One can imagine that it was built at the L-5 shipyard to a 90% complete condition: no shielding; no soil in the farm modules; and, very little water onboard so as to keep its mass to a minimum for the transit to Mars. It will be completed once it reaches high Mars orbit using materials from the Martian moons. Its purpose there will be to serve as the home base for the people who are working to terraform Mars, people who have the dream, the vision, and the determination to start the process of making Red Mars into Green Mars.

With a clear and realistic vision of the future, it becomes possible to move mountains, build castles in the sky, and remake worlds.

Author

Dare greatly.

Learn from your mistakes.

Don't stop.

Chapter 23: The Final Ingredient

It takes courage to live life fully.

It takes courage to open your heart to someone and really love them. Especially when you know how much it can hurt when it doesn't work.
It takes courage to start a family.
It takes courage to start a business.
It takes courage to go away to school and learn a profession.
It takes courage to have dreams and to pursue them.
It took courage for our ancestors to leave their homes in search of a better life in a new place.
It took courage to build ships and transcontinental railroads and use them.
Everything new takes courage. Everything new requires a leap of faith into the unknown.

All our progress as individuals, as a species, as a civilization, has come from people having the courage to risk it all in search of a better life for themselves and their children. Every time it happens on a mass level, it separates those who have the courage to grow and change from those who do not.

When our ancestors began leaving Africa 60,000 years ago it separated those with courage and a desire for growth from those who did not. It happens every time another migration occurs. The most recent mass migration was from Western Europe to North America. The end result was the creation of the greatest country the world has ever known, the United States of America.

There is another mass migration getting ready to occur that will be more profound than any that has occurred before. It will be our movement into space, to Space Colonies, to the asteroids, to the Moon and the moons of other planets, and to Mars. It will be the greatest challenge and the greatest adventure that the human race has taken on to date. It will separate people into those who have the courage, and those who do not.

Look at our history and notice how the leading edge of our progress and our continuing evolution as a species takes place on those new frontiers. It does not occur with those who stay behind. New discoveries are made on the frontier. New technologies are developed on the frontier. Incredible fortunes are made on the frontier. With great risk comes great reward.

It takes courage to live life fully.

Every mass migration has distilled out the best, the brightest, the strongest, the most ambitious, those with a dream for a better future, and all with the courage to make it happen. It is evolution in action because of the separation and concentration of our best in a new environment, and because they have chosen the difficult path.

As former President Kennedy said,

We choose to go to the Moon in this decade and do the other things, not because they are easy, but because they are hard.

As with any major societal event, there will be those who speak against it. It doesn't matter. Based on our history it will happen in spite of the naysayers. It always has. At their worst, the most the naysayers can do is to slow it down, they cannot stop it. The only thing that can stop it will be a failure of courage on the part of those who want to see it happen.

To make this clear, there are 350 million people in the United States of America, of which 120 million have jobs. The average income for each of those jobs is approximately $50,000 per year or $200 per day. Polls have shown that approximately 25% of Americans are avid supporters of space exploration. That works out to 30 million jobholders. If just half of those 30 million jobholders invested one day's wages in order to make this happen, that would raise $3 billion. That is more than enough to get this started. Quite simply, it does not have to be a monstrous government project. Those who have the dream and who want to make it happen can fund it with minimum risk.

As shown in previous chapters there are a number of possible ways to build this bridge into space. As just shown, it is possible for those who want to see this happen to make it happen. The only remaining question is, "do we have the courage, the will, and the determination to do it?"

If you are young and looking for a dream to dedicate your life to that fills you with excitement and passion for life, I hope you will consider this dream for our future in space. I hope you will educate yourself about it, save your money for it, and become part of it. If you are older, I hope you will support it, invest in it, and encourage your children and any other young people you know to get involved in it.

Two hundred years ago, in the United States of America, our ancestors believed that it was their manifest destiny to grow the United States of America until it reached across the North American continent from the East Coast to the West Coast. Guided by that vision they made it happen.

Today it is my belief that it is our manifest destiny to become a spacefaring civilization. To make that happen we need to build a mass-market affordable transportation system that will open up the solar system for development and settlement. It will be the modern-day equivalent of a transcontinental railroad into space. It is a railroad that will give us the solar system. I hope you will find the courage in yourself to be part of it.

It takes courage to live life fully.
It takes courage to work at being your very best.
It takes courage to make your dreams come true.
It takes courage to be explorers and build a spacefaring civilization.
It is time for us to be great again.

Epilogue

Everyone's a star and deserves the right to twinkle.

Marilyn Monroe

This photograph is a compilation of long duration exposures taken by the Hubble Space Telescope over a period of years. All of these exposures are of a small region of space in the constellation Fornax, which is located southwest of the constellation Orion. It is an area of space that is approximately one-tenth the diameter of a full Moon in width and height. This particular patch of sky was chosen for its lack of visible stars, which makes it easier to see the stars that are really faint and far away. If you were to look at this particular patch of sky on a dark clear night you would see nothing but blackness.

It is estimated that this picture of this little patch of blackness contains over 10,000 galaxies.

Our galaxy, the Milky Way Galaxy, is thought to contain 100 billion stars.

It would take 13 million photographs like this one to cover the entire sky.

This is the universe we live in.

It really is that big.

Considering the overall size of everything, it makes the idea of building carrier aircraft, ground accelerators, reusable rockets and a Skyhook, no big deal at all.

So how about we get on with it?
Carrier aircraft for air-launching rocket planes have been around for years.
Ground accelerators have been around for years.
SpaceX and Blue Origin are building reusable first stage rockets.
Space stations have been around for years.
Ion propulsion systems have been around for years.
So why not attach a long cable to a Space Station and combine all these ideas together and make spaceflight affordable?

There is a mighty big universe out there and it is calling.

Can you hear it?

Hubble Space Telescope, Extreme Deep Field, 2014

An Orion spacecraft in the process of being released from the Upper Endpoint Station and heading for higher orbit. *By Nick Stevens*

About the Author

I like to design and build things. It is my passion. I especially like designing and building things that move and that have never been thought of before. I have designed, built, and test flown hang gliders, ultralight aircraft, and homebuilt aircraft. I have helped design and build high-performance aircraft such as the F-5, F-18, F-20, F-22, F-35, U-2, SR-71, and the F-117. I have helped design satellites, space stations, and launch vehicles. Some of my work is in orbit around the Earth and my name is on a satellite that is headed out of the solar system. I have done original research and have invented wondrous things. I am the person who invented the non-rotating Skyhook. I have multiple degrees and I am a licensed ship's officer.

I have sailed all around the world on 600 plus foot long ships that weighed in excess of 40,000 tons. I have sailed in the Atlantic, the Pacific, and the Indian Ocean. I have sailed all the major seas and many of the less well known. I have sailed to South America, Central America, the South Pacific, the Far East, Europe, the Baltic, Africa, Asia, and the Middle East.

I have experienced great passions, great joys, great enthusiasms, and magical wonders. I have experienced success and failure, victory and defeat. I have also experienced great love and tragic loss.

I believe that it isn't what happens to us that matters but how we respond to it that does. My primary motivator is to be more than I now am.

If you wonder about my qualifications for writing this book: I am qualified. If that isn't enough for you, read the references. The answers are all there.

Somebody once asked me, "what makes you so special?"

I am not special. I am simply one of 7 billion people on planet Earth, in a galaxy of 100 billion stars, in a universe with galaxies like grains of sand. In the overall scheme of things, all of us are utterly insignificant. Yet it does not matter to the universe that we are insignificant, the universe shares its secrets with anyone who is willing to observe, study, experiment, and learn.

This book is the result of many years of study, experimentation, and learning. It is both a gift to the future to those who want to build a better world, and a thank you to those who have contributed in a positive way to the world we enjoy today.

References

1 Wernher von Braun, The Mars Project, University of Illinois Press 1953

2 Wernher von Braun and others, Across the Space Frontier, The Viking Press 1952

3 Wernher von Braun and others, Conquest of the Moon, The Viking Press 1953

4 Wernher von Braun and others, The Exploration of Mars, The Viking Press 1956

5 Wernher von Braun and others, Man Will Conquer Space Soon! Published in Collier's Magazine between 1952-54. http://www.aiaahouston.org/Horizons/Horizons_2013_09_and_10.pdf

6 Walt Disney, Wernher von Braun and others, Man in Space, Disneyland episode 1955

7 Walt Disney, Wernher von Braun and others, Man and the Moon, Disneyland episode 1955

8 Walt Disney, Wernher von Braun and others, Mars and Beyond, Disneyland episode 1957

9 O'Neill, G., The High Frontier, William Morrow and Company 1977

10 Hadingham, E., Ancient Chinese Explorers, NOVA episode, PBS 2001

11 Malthus, T. R., An Essay on the Principle of Population, 1798, Oxford World's Classic reprint http://www.esp.org/books/malthus/population/malthus.pdf

12 Dvorsky, G. (13 February 2013). Why we'll probably never build a space elevator. io9.com. http://io9.com/5984371/why-well-probably-never-build-a-space-elevator

13 Feltman, R. (7 March 2013). Why Don't We Have Space Elevators? . Popular Mechanics. http://www.popularmechanics.com/space/a8814/why-dont-we-have-space-elevators-15185070/

14 Scharr, Jillian (29 May 2013). Space Elevators On Hold At Least Until Stronger Materials Are Available, Experts Say. Huffington Post. http://www.huffingtonpost.com/2013/05/29/space-elevators-stronger-materials_n_3353697.html

15 Templeton, Graham (6 March 2014). 60,000 miles up: Space elevator could be built by 2035, says new study. Extreme Tech. Retrieved 2014-04-19. http://www.extremetech.com/extreme/176625-60000-miles-up-geostationary-space-elevator-could-be-built-by-2035-says-new-study

16 Darnell (10 April 2009). Are Traditional Space Elevators the Wrong Way Up? . Colony Worlds. Archived from the original on 2009-07-14. http://colonyworlds.com/2009/04/are-traditional-space-elevators-the-wrong-way-up.html

17 Mankins, J.C., The MagLifter: An Advanced Concept Using Electromagnetic Propulsion in Reducing the Cost of Space Launch, AIAA 94-2726, 30th Joint Propulsion Conference June 27-29 1994. https://upload.wikimedia.org/wikipedia/commons/5/58/Maglifter_Mankins.pdf

18 Durazo, F., Vertical Accelerator Launch Tower, https://valtrockets.wordpress.com

19 Meyer, C., Sky Ramp Technology, http://www.g2mil.com/skyramp.htm

20 Thomas, S.R., Performance Evaluation of the NASA GTX RBCC Flowpath. NASA/TM-2001-210953. http://ntrs.nasa.gov/archive/nasa/casi.ntrs.nasa.gov/20010092480.pdf

21 Olds, J.R., Argus, a Highly Reusable SSTO Rocket-Based Combined Cycle Launch Vehicle with MagLifter Launch Assist, NASA/CR-1998-208093. http://ntrs.nasa.gov/archive/nasa/casi.ntrs.nasa.gov/19980202962.pdf

22 NASA X-43 Demonstrates Hypersonic Flight with a Scramjet,

http://www.nasa.gov/centers/armstrong/news/FactSheets/FS-040-DFRC.html

23 Sarmont, E. (26 May 1990). An Orbiting Skyhook: Affordable Access to Space. International Space Development Conference (Anaheim California). http://archivist.nss.org/Public/ISDC/ISDC1990_Anaheim/ConferenceMisc-DaleAmonPapers/19900525-19900528-Anaheim1990-ISDC9-Program/19900525-19900528-Anaheim1990-ISDC9-Program-p006-007.jpeg

24 Sarmont, E. (October 1994). How an Earth Orbiting Tether Makes Possible an Affordable Earth-Moon Space Transportation System. SAE 942120. http://papers.sae.org/942120/

25 Zubrin, R. (March 1995). The Hypersonic Skyhook. Journal of the British Interplanetary Society **48** (3): 123–128. Bibcode:1995JBIS...48..123Z. http://adsabs.harvard.edu/abs/1995JBIS...48..123Z

26 Cosmo, M.; Lorenzini, E. (December 1997). Tethers in Space Handbook (PDF) (Third ed.). Smithsonian Astrophysical Observatory. Archived (PDF) from the original on 2003-03-22. http://www.tethers.com/papers/tethersinspace.pdf

27 Sarmont, E., Affordable to the Individual Spaceflight. Archived from the original on 2007-02-13. http://web.archive.org/web/20070213093349/http://www.affordablespaceflight.com/home.html

28 Wilson, N. (August 1998). Space Elevators, Space Hotels and Space Tourism. SpaceFuture.com. http://www.spacefuture.com/archive/space_elevators_space_hotels_and_space_tourism.shtml

29 Smitherman, D.V., Space Elevators, An Advanced Earth-Space Infrastructure for the New Millennium. NASA/CP-2000-210429. Archived from the original on 2007-02-21. http://www.nss.org/resources/library/spaceelevator/2000-SpaceElevator-NASA-CP210429.pdf

30 Mottinger, T.; Marshall, L., The Bridge to Space – A space access architecture. AIAA 2000-5138. http://arc.aiaa.org/doi/abs/10.2514/6.2000-5138

31 Mottinger, T.; Marshall, L. (2001). The Bridge to Space Launch System. Space Technology and Applications International Forum. CP552. https://www.deepdyve.com/lp/american-institute-of-physics/the-bridge-to-space-launch-system-NqveWCXQ2e

32 Marshall, L.; Ladner, D.; McCandless, B. (2002). The Bridge to Space: Elevator Sizing & Performance Analysis. Space Technology and Applications International Forum. CP608. https://www.deepdyve.com/lp/american-institute-of-physics/the-bridge-to-space-elevator-sizing-performance-analysis-lozS43VU5q

33 Stasko, S; Flandro, G., The Feasibility of an Earth Orbiting Tether Propulsion System. AIAA 2004-3901. http://arc.aiaa.org/doi/abs/10.2514/6.2004-3901

34 Newton, K.; Canfield, S. (6 July 2005). NASA Engineers, Tennessee College Students Successfully Demonstrate Catch Mechanism for Future Space Tether. News Release (Marshall Space Flight Center). Archived from the original on 2005-09-08. http://web.archive.org/web/20050908101527/http://www.nasa.gov/centers/marshall/news/news/releases/2005/05-108.html

35 Rogers, L. (21 March 2008). It's ONLY Rocket Science. Springer Science and Business Media. p. 298. http://books.google.com/books?id=75b84eC-ulsC&pg=PA298&lpg=PA298&dq=orbital+skyhooks&source=bl&ots=cgNjxCPNxy&sig=_rPiooo_0g12up9Y6L_PurVsc4s&hl=en&sa=X&ei=F1O4U6jrHZK3yASy1oGABQ&ved=0CC8Q6AEwAzgU#v=onepage&q=orbital%20skyhooks&f=false

36 Andrews, D. (3 September 2009). Advanced ETO Space Transportation. NASA Langley Advanced Space Transportation Workshop. Archived from the original on 2014-07-13. http://web.archive.org/web/20140713153047/http://niacms.nianet.org/resources/Education/Continuing-Education/Workshops/Workshops/pdfs/Andrews_Next-Gen-RLVs2.aspx

37 Andrews, D., Space Colonization: A Study of Supply and Demand (PDF). Archived (PDF) from the original on 2012-03-11. http://web.archive.org/web/20120311204915/http://www.spacearchitect.org/pubs/IAC-11-E5.1.8.pdf

38 Chen, Yi; Huang, Rui; Ren, Xianlin; He, Liping; He, Ye (2013). History of the Tether Concept and Tether Missions: A Review. ISRN Astronomy and Astrophysics **2013**. doi:10.1155/2013/502973. Retrieved 2014-03-07. http://www.hindawi.com/journals/isrn/2013/502973/

39 Sarmont, E., Affordable Access to Space: Basic Non-Rotating Skyhook with Falcon 9 & Dragon. https://www.dropbox.com/s/fp6q6eq9iky8vwo/Affordable%20Access%20to%20Space%2C%201.pdf

40 TORAY Industries Carbon Fiber Data Sheet. http://www.torayca.com/en/index.html

41 Penzo, P. (July 1984). Tethers for Mars Space Operations. The Case for Mars II. 62 Science and Technology Series. pp. 445–465.

42 Penzo, P.; Carroll, J. (December 1997). Mars Moons Tether Transport System (PDF). Tethers in Space Handbook (3rd ed.). pp. 70–71. Archived (PDF) from the original on 2003-03-22. http://web.archive.org/web/20030322134829/http://www.tethers.com/papers/TethersInSpace.pdf

43 Weinstein, L., Space Colonization Using Space-Elevators from Phobos. http://www.nss.org/settlement/mars/2003-SpaceColonizationUsingSpaceElevatorsFromPhobos.pdf

44 Moravec, H.P. (18–20 October 1977). A Non-Synchronous Orbital Skyhook. 23rd AIAA Meeting - The Industrialization of Space (San Francisco, CA). Archived from the original on 1999-10-09. http://web.archive.org/web/19991009023915/http://www.frc.ri.cmu.edu/~hpm/project.archive/1976.skyhook/papers/scable.pox

45 Moravec, H. (1978). Non-Synchronous Orbital Skyhooks for the Moon and Mars with Conventional Materials. Archived from the original on 1999-02-03. http://web.archive.org/web/19990203022459/http://www.frc.ri.cmu.edu/%7Ehpm/project.archive/1976.skyhook/papers/scasci.txt

46 Moravec, H. (1978). Free Space Skyhooks. Pittsburgh Pa: The Robotics Institute Carnegie-Mellon University. Archived from the original on 1999-10-09. http://web.archive.org/web/19991009012319/http://www.frc.ri.cmu.edu/~hpm/project.archive/1976.skyhook/papers/free.txt

47 Moravec, H. (1986). Orbital Bridges. Pittsburgh Pa: The Robotics Institute Carnegie-Mellon University. Archived from the original on 1999-10-12. http://web.archive.org/web/19991012232419/http://www.frc.ri.cmu.edu/~hpm/project.archive/general.articles/1987/skyhook.ltx

48 Baumann, E.; Pado, L.E. (1999). An Interplanetary Mass Transit System Based on Hypersonic Skyhooks and Magsail-Driven Cyclers (PDF). Gateway Space Transport Inc. Archived (PDF) from the original on 2012-08-17. http://www.marspapers.org/papers/Baumann_1999.pdf

49 Bogar, Thomas J.; Bangham, Michal E.; Forward, Robert L.; Lewis, Mark J. (7 January 2000). Hypersonic Airplane Space Tether Orbital Launch System. Research Grant No. 07600-018l Phase I Final Report (PDF). NASA Institute for Advanced Concepts. Archived (PDF) from the original on 2005-12-18. Retrieved 2014-03-20. http://web.archive.org/web/20051218225817/http://www.niac.usra.edu/files/studies/final_report/355Bogar.pdf

50 Tether Launch Assist. Tethers Unlimited Inc. Archived from the original on 2003-02-07. http://web.archive.org/web/20030207055339/http://www.tethers.com/LaunchAssist.html

51 Space Tether Bibliography. Tethers Unlimited Inc. Archived from the original on 2014-07-13. Retrieved 4 July 2014. http://www.tethers.com/bibliography.html

52 Levin, E.M. (2007). Dynamic Analysis of Space Tether Missions. Advances in the Astronautical Sciences (American Astronautical Society) **126**.
http://books.google.com/books?id=ZRiJKGbxWBQC&pg=PA237&lpg=PA237&dq=non-synchronous+skyhook&source=bl&ots=Sjxy2-GAJo&sig=XUZ4gJtOdsG0wWidVC7XzBtJn8g&hl=en&sa=X&ei=g4-0U_fgJY62yAT9goHABQ&ved=0CB4Q6AEwATgK#v=onepage&q=non-synchronous%20skyhook&f=false

53 Stuhlinger, E., Possibilities of Electric Space Ship Propulsion, 5th International Astronautical Congress, Innsbruck, Austria, August 5-7, 1954, pp 100-119

54 Stuhlinger, E., Electrical Propulsion System for Space Ships with Nuclear Power, NASA TMX-57089, July 1, 1955
http://www.astronautix.com/craft/stus1957.htm

55 Stuhlinger, E., and King, J., Concept for a Manned Mars Expedition with Electrically Propelled Vehicles, Progress in Astronautics Vol. 9, 1963, pp 647-664
http://www.astronautix.com/craft/stus1962.htm

56 Stevens, N., Ernst Stuhlinger's Sun Ship, International Space Art Network, video [3]
https://www.youtube.com/watch?v=-hHVgdvzCj4

57 Stevens, N., Ernst Stuhlinger's Umbrella Ship, International Space Art Network, video [4]
https://www.youtube.com/watch?v=UenbTst0ZZU

58 The Incredible Ions of Space Propulsion. Science at NASA. NASA. 16 June 2000. Archived from the original on 2011-10-23.
http://web.archive.org/web/20111023082405/http://science1.nasa.gov/science-news/science-at-nasa/2000/ast15jun_1/

59 Sovey, J.S.; Rawlin, V.K.; Patterson, M.J. (May–June 2001). Ion Propulsion Development Projects in U.S.: Space Electric Rocket Test 1 to Deep Space 1. Journal of Propulsion and Power **17** (3): 517–526.
https://archive.org/stream/nasa_techdoc_20010093217/20010093217#page/n0/mode/2up

60 Nautilus-X: A Real Spaceship At Last, video by Fragomatik, 2014.
https://www.youtube.com/watch?v=hObbL4DCesI

61 Keaton, P., A Moon Base/Mars Base Transportation Depot, Lunar Bases and Space Activities of the 21st Century, National Academy of Sciences, Washington, DC, Oct 29-31, 1984 [5]
http://articles.adsabs.harvard.edu/cgi-bin/nph-iarticle_query?bibcode=1985lbsa.conf..141K

62 Stevens, N., Ernst Stuhlinger's Mars Ion Rocket, video
https://search.yahoo.com/search;_ylc=X3oDMTFiN25laTRvBF9TAzIwMjM1MzgwNzUEaXRjAzEEc2VjA3NyY2hfcWEEc2xrA3NyY2h3ZWI-?p=nick+stevens+ion+rocket&fr=yfp-t-201&fp=1&toggle=1&cop=mss&ei=UTF-8

63 Stevens, N., Opening the High Frontier, video
https://vimeo.com/175424301

Appendix 1: Launch Vehicles

Titan 2 GLV

1st Stage	
empty	6,736 kg
propellant	111,130 kg
stage total	117,866 kg
Isp	296 seconds
2nd Stage	
empty	2,404 kg
propellant	26,535 kg
stage total	28,939 kg
Isp	316 seconds
Gemini Spacecraft	3,600 kg
Total Mass	150,405 kg

Apparent Velocity	
1st Stage	3,895 m/s
2nd Stage	5,234 m/s
Total	9,129 m/s

Payload Fraction	2.4 %
Empty Wt. Fraction	6.1 %
Propellant Fraction	91.5 %
Propellant/Payload	38.24
Vehicle/Payload	2.54
Prop. & Vehicle/Payload	40.78

NASA photo

Appendix 1: Launch Vehicles

Saturn 1B

1st Stage	
empty	49,250 kg
propellant	399,400 kg
stage total	448,650 kg
Isp	296 seconds
2nd Stage	
empty	15,200 kg
propellant	103,600 kg
stage total	118,800 kg
Isp	421 seconds
Lunar Module Adaptor (LMA)	1,840 kg
Command and Service Module (CSM)	16,280 kg
Launch Escape System (LES)	4,200 kg
Total Mass	589,770 kg
Apparent Velocity	
1st Stage	3,280 m/s
2nd Stage	5,830 m/s
Total	9,110 m/s
Payload Fraction	2.76 %
Final Wt. Fraction	11.95 %
Propellant Fraction	85.29 %
Propellant/Payload	30.90
Vehicle/Payload	4.33
Prop. & Vehicle/Payload	35.23

NASA photo

Saturn V

1st Stage	
empty	130,000 kg
propellant	2,160,000 kg
stage total	2,290,000 kg
Isp	304 seconds

2nd Stage	
empty	40,100 kg
propellant	456,100 kg
stage total	496,200 kg
Isp	421 seconds

3rd Stage	
empty	13,300 kg
propellant to LEO	22,440 kg
prop. LEO to LTO	84,160 kg
stage total	119,900 kg
Isp	421 seconds

LMA	1,840 kg
LEM	33,500 kg
CSM	28,800 kg
LES	4,200 kg
Total Mass	2,974,440 kg

Apparent Velocity	
1st Stage	3,859 m/s
2nd Stage	4,580 m/s
3rd Stage to LEO	536 m/s
Total to LEO	8,975 m/s
Speed of LEO	7,800 m/s
LEO to LTO	3,035 m/s
Speed of LTO	10,835 m/s
Total	12,010 m/s

	LEO	LTO
Payload, kg	161,600	62,300
Payload Fraction	5.4%	2.1%
Empty Wt. Fraction	5.9%	6.4%
Propellant Fraction	88.7%	91.5%
Propellant/Payload	16.3	43.7
Vehicle/Payload	1.1	3.0
P+V/Payload	17.4	46.7

NASA photo

Appendix 1: Launch Vehicles

Space Shuttle

Solid Rocket Boosters (2)	
empty	91,000 kg
propellant	499,000 kg
total	590,000 kg
Isp	268 seconds
burn time	127 seconds
External Tank	
empty	26,500 kg
propellant	729,500 kg
total	756,000 kg
Space Shuttle	75,000 kg
Payload, LEO	24,000 kg
OMS propellant	5,000 kg
Total Mass	2,040,000 kg
Propellant Burned	
Isp, SRB+SSME	314 seconds
propellant, SRB+ET	1,192,700 kg
Isp, SSME	455 seconds
propellant, ET	534,800 kg
Isp, OMS	316 seconds
propellant, OMS	2,600 kg
Apparent Velocity	
SRBs + SSMEs	2,704 m/s
SSMEs	7,263 m/s
OMS	78 m/s
Total	10,045 m/s
Payload Fraction	1.2 %
Empty Wt. Fraction	13.9 %
Propellant Fraction	84.9 %
Propellant/Payload	72.1
Vehicle/Payload	1.1
Prop. & Vehicle/Payload	73.2

NASA photo

Appendix 1: Launch Vehicles

NASA photo

Pegasus The Pegasus is an air-launched launch vehicle that was developed by Orbital Sciences Corp. in the late 1980s. Its first flight was in 1990.

1st Stage	
empty	1,868 kg
propellant	12,152 kg
stage total	14,020 kg
Isp	294 seconds
2nd Stage	
empty	345 kg
propellant	3,025 kg
stage total	3,370 kg
Isp	292 seconds
3rd Stage	
empty	203 kg
propellant	782 kg
stage total	985 kg
Isp	293 seconds
Payload	375 kg

Total Mass	18,750 kg
Apparent Velocity	
1st stage	3,009 m/s
2nd stage	2,920 m/s
3rd stage	2,457 m/s
Total	8,386 m/s
Payload Fraction	2.0 %
Empty Wt. Fraction	12.9 %
Propellant Fraction	85.1 %
Propellant/Payload	42.5
Vehicle/Payload	6.5
Prop. & Vehicle/Payload	49.0

Appendix 2: The Rocket Equation

This appendix is written using basic algebra. It is written for young people and those who are young at heart, who are interested in learning the basic equations of being a rocket scientist. All that is needed is a willingness to learn and a scientific calculator that has a button with e on it. This button is for natural logarithms, which is a way of counting using exponents. If you want to learn more about natural logarithms all you have to do is Google 'natural logarithm' and there are plenty of good articles that will tell you all about it. Otherwise, all you need to do is push the e button whenever you see the e symbol in an equation.

You can also put these equations in an Excel Spreadsheet on your home computer and use them to make graphs that will show the propellant fraction for a launch vehicle for a wide range of velocities and different specific impulse values.

If you are in school and are having trouble with the math or using the spreadsheet, go talk to your math teacher. Odds are they will be delighted to assist you. Remember, the only dumb question is the one you don't ask.

If you are interested in the history of the rocket equation, it was invented by Konstantin Tsiolkovsky. That is another good name to Google.

The Rocket Equation

$$\frac{m_i}{m_f} = e^{\frac{\Delta v}{g * Isp}}$$

where:

m_i = mass initial, also know as the mass of the launch vehicle at lift off

m_f = mass final, also known as the mass of the launch vehicle after the usable propellant is consumed

m_{prop} = mass of the usable propellant

$m_f + m_{prop} = m_i$ or $m_i - m_{prop} = m_f$

e = natural logarithm

Δv = delta v, also known as change in velocity, or the velocity required to get where you want to go

g = acceleration of gravity = 9.8 meters/second^2

Isp = specific impulse, the numbers of seconds a rocket engine will run for every pound of propellant, for every pound of thrust

Examples of specific impulse are:

- A rocket engine that uses liquid oxygen and liquid hydrogen for propellant has a specific impulse of approximately 460 seconds in a vacuum.
- A rocket engine that uses liquid oxygen and kerosene for propellant has a specific impulse of somewhere between 300 seconds and 340 seconds (depends on the rocket motor) when operating in a vacuum.

An example of how to use the rocket equation

The year is 2025 and you are vacationing in the American southwest. While there you stop in at Spaceport America where you get to see SpaceShipTwo make a flight into suborbital space. You also get to see 'Big Bob's Boosters' a fairly new start-up company that provides an air-launched reusable first stage launch service for spacecraft and payloads that are headed to orbit and to the Skyhook that orbits the Earth. It is an interesting tour.

As you leave the spaceport you see a small sign on the side of the highway that says 'Cowboy Cal's Used Spaceships' with an arrow that points to a small side road. On a whim, you decide to check it out. As it turns out Cowboy Cal's Used Spaceships is more of a space age junkyard than anything else, but in the back of the lot, you find an old Gemini capsule that is sealed up in an airtight shipping container as well as the upper stage of a Titan 2 launch vehicle. Cowboy Cal sees your interest and stops by to say "Howdy." It turns out the Gemini capsule was 99% finished when the Gemini program ended and was sold as surplus to one of the engineers who worked on the project. The engineer finished building it in his garage and then sealed it up in the shipping container in a nitrogen atmosphere where it has been ever since. Cowboy Cal bought it from the estate of the engineer's widow a few years ago. "I'll sell it to you for my cost plus 10% if you're interested," he says.

Your brain goes into overdrive as your mind starts thinking about 'Big Bob's Boosters' that accelerates upper stages and their payloads to a speed of 2,100 meters per second and the Skyhook whose lower endpoint is moving 500 meters per second slower than orbital velocity for its altitude. Then there is that Titan 2 upper stage; does it have enough juice to make up the difference?

You get out your cell phone and start researching the details on the Internet. The Titan 2 upper stage has an empty mass of 2,404 kilograms, a take-off mass of 28,939 kilograms, and a specific impulse of 316 seconds. The Gemini spacecraft had a take-off mass of 3,600 kilograms, which included two astronauts in spacesuits plus all their gear and consumables. That comes out to a mass initial of 3,600 + 28,939 = 32,539 kilograms. Mass final after all the usable propellant is burned is 3,600 + 2,404 = 6,004 kilograms. Mass initial divided by mass final is 5.42.

Now for the other half of the equation: the lower end of the Skyhook is 185 kilometers up. The speed for orbit at that altitude is 7,800 meters per second. The lower end of the Skyhook is moving 500 meters per second slower than that. That means I will need to be moving at 7,300 meters per second if I am going to rendezvous with the lower end of the Skyhook. Big Bob's Boosters will get me up to 2,100 meters per second out of the atmosphere. That means the Titan 2 upper stage will need to accelerate me to 5,200 meters per second if this is going to work.

First I will need to calculate the exponent term for *e.*

$$\frac{\Delta v}{g * Isp} = \frac{5200}{9.8 * 316} = 1.679$$

Now use that as the exponent for *e.*

$$e^{1.679} = 5.36$$

Mass initial divided by mass final, is also known as the mass ratio. As already calculated the mass ration for the Gemini with Titan 2 upper stage is 5.42. The mass ratio required to get to the lower end of the Skyhook is 5.36. The upper stage has a little more mass ratio than need. This will work.

I look up from my calculations and Cowboy Cal has a smile on his face and a twinkle in his eyes. "You a 'real' pilot?" he asks. I nod yes. "Thinking of making a flight? Maybe do a little star walking?" he asks.

I say, "The thought has crossed my mind."

His smile gets even bigger; "Dreams of space are hard to ignore," he says. "I'll give you a real good deal on the upper stage if you'll let me paint 'Cowboy Cal's Used Spaceships' on the side of it. I can also get you a deal on a new 3D printed rocket motor for it if you want. My nephew, Billy Bob, runs that outfit 'Big Bob's Boosters' across the highway. I also got some spacesuits in that locker over there. I am sure there is one that will fit you."

Propellant Fraction

To convert the answer you got from the Rocket Equation into the propellant fraction, just plug the 5.36 from above into the following equation as shown.

$$propellant\ fraction = 100 * \left[1 - \frac{1}{5.36}\right] = 81.34\%$$

You can double-check this answer by dividing the propellant mass of the upper stage you just bought from Cowboy Cal by the total take-off mass and then multiply that answer by 100.

upper stage take-off mass	28,939 kg
minus upper stage empty mass	2,404 kg
gives a propellant mass of	26,535 kg

Now divide that answer by the total take-off mass of the Gemini spacecraft and upper stage, and multiply that answer by 100.

propellant mass	26,535 kg
divided by the total vehicle mass	32,539 kg
equals	.81548
now multiply that by 100	81.548

That comes out to a propellant fraction of 81.55%.

The difference between the two answers is that the first one, the 81.34%, is the amount of usable propellant that is needed to get to the lower end of the Skyhook. The second one, the 81.55%, is what the vehicle actually carries.

Made in United States
Orlando, FL
04 November 2026

90204422R00098